Facilities Management Handbook

Facilities Management Handbook

John Molnar, P. E.

VNR VAN NOSTRAND REINHOLD COMPANY
NEW YORK CINCINNATI TORONTO LONDON MELBOURNE

Library of Congress Catalog Card Number: 82-24768
ISBN: 0-442-26347-3

Manufactured in the United States of America

Published by Van Nostrand Reinhold Company Inc.
135 West 50th Street, New York, N.Y. 10020

Van Nostrand Reinhold
480 Latrobe Street
Melbourne, Victoria 3000, Australia

Van Nostrand Reinhold Company Limited
Molly Millars Lane
Wokingham, Berkshire, England

15 14 13 12 11 10 9 8 7 6 5 4 3 2 1

Library of Congress Cataloging in Publication Data

Molnar, John.
Facilities management handbook.

Bibliography: p.
Includes index.
1. Building—Handbooks, manuals, etc. 2. Building management—Handbooks, manuals, etc. I. Title.
TH151.M59 1983 690′.068 82-24768
ISBN 0-442-26347-3

To my wife
Elaine Naden Molnar
and our children
Susan and Jay

Preface

Facilities Management Handbook was written from practical experience to consolidate, under one cover, all the necessary information at an adequate depth to guide you effectively through the intricacies of a project that may begin with site search, progress through leasing, new building construction or remodeling, and on to occupancy. This is not a theoretical exposition, but instead is a practical approach based on 30 years experience with every aspect of the material covered. These methods and concepts have been successfully used in actual situations.

The book's purpose is to bring together, in one handy volume, information usually found in separate, specialized, technical publications, in an easy-to-read style readily comprehensible and usable by both technical and nontechnical people. It was written to serve anyone responsible for building design and construction, facilities management and operations, and real estate leasing; particularly building owners and managers, industrial, commercial, and institutional facilities department personnel, plant engineering, and real estate departments. It could also be valuable to students and others planning careers in these fields. The book provides necessary information to assist sales personnel handling products and services serving the need of the above.

To obtain the greatest benefit from this book, it is important to become familiar with its contents. At first reading, the book should be skimmed from cover to cover for familiarization with the method and sequence of material presented, and to get a feel for the author's technique. Then proceed to the chapter or chapters that address your specific problem. While the chapters are organized sequentially to follow a normal progression starting with site search on through occupancy, chapters may be used individually to address specific problems, or may be combined for broad scope projects. For example, chapter 1 deals with site search and selection. This may be considered as a stand-alone chapter to deal with that particular problem. But if the project also includes new building construction on that site, other chapters are included that address additional necessary functions, such as, selecting design consultants, cost estimating techniques, construction administration, mechanical systems and equipment, electrical systems and equipment, and move and occupancy. Depending upon the scope of the project and the degree of your participation, you will find the necessary information in one or more chapters as required.

This book contains numerous checklists, developed by the author, designed to serve as guides in dealing with problems. Checklists are useful because they provide a written list of items that must be addressed and to eliminate the probability of oversight or omission. A typical commercial lease is included in chapter 2, and it is essential reading if you are involved in a leasing situation. Chapters 9 and 10 include typical specifications to illustrate the manner in which specifications are written, and to describe essential items that must be included. These may be used as a guide to write your own specifications when necessary or to review specifications written by consultants. Specifications and drawings are contract documents that form the basis for an agreement between an owner and contractor and should be treated accordingly.

Last, and probably most important, are the chapter summaries. They were written in the checklist technique to provide a simple step-by-step summary of the material presented or problem-solving procedure which was applicable.

In this book, like in a typical lease agreement from which it was excerpted, the words of any gender shall mean to include any other gender.

JOHN MOLNAR

Acknowledgments

In a book of this type the author depends considerably upon the generosity of manufacturers, associations, and societies for their cooperation in supplying information on the latest techniques, procedures, equipment, and systems. The following are the names of those that have supplied data and photographs, and have assisted in other ways:

- Buffalo Forge Company, Buffalo, NY
- Crane Co. New York, NY
- Gould Inc., Electric Fuse Division, Newburyport, MA
- Gould Inc., Electric Motor Division, St. Louis, MO
- ITT, Fluid Handling Division, Morton Grove, IL
- Jenkins Bros., Norwalk, CT
- Sprague Electric Company, North Adams, MA
- The Trane Company, LaCrosse, WI
- American Society of Heating, Refrigerating and Air-Conditioning Engineers, Inc., Atlanta, GA
- Building Owners and Managers Association International, Washington, DC

Some of the material in this book has appeared in some form, written by the author, in the trade magazines listed below.

"Selecting the Right Site," *Area Development,* June, 1974.

"Real Estate & Leasing Procedure," *Area Development,* March, 1974.

"How to Compare Office Rental Costs," *Area Development,* November, 1974.

"The Role of Consulting Services in Facilities Planning," *Area Development,* December, 1977.

"How to Zero in on Specific Site," *Area Development,* April, 1980.

"Determining Plant Office Space Requirements," *Plant Engineering,* May 16, 1974.

"Cost Estimating Techniques," *Building Operating Management,* March, 1975.

"Space Planning," *Building Operating Management,* June, 1976.

"Remodeling Decision Making Process," *Building Operating Management,* August, 1979.

Contents

Facilities Management Handbook

1. Site Search and Selection

There comes a time in the life of a company or corporation when site search and selection is necessary. For some companies this is a routine occurrence, and it is done on a regular basis using qualified and experienced personnel. For others, this may happen once in a life time, and the company may have no concept where to begin or how to proceed. The vast majority of companies fall in between these two extremes. Considering the broad range of skills avilable within the various companies, the problem of site search and selection is further complicated by the people with available sites. This is evidenced by the extensive advertising by the smallest communities to the largest cities, from small industrial/commercial park developers to huge new city complexes, by speculative builders, regional planners, realtors, and professional firms that specialize in planning for and conducting the site search. The inducement offered to the prospective client companies is as varied as is the type and location of the sites.

Real estate frequently represents a major portion of a company's capital investment. Sound real estate investment decisions could have a significant impact on a company's degree of success and growth. Dollars misspent on real estate deprive a company of investment resources for materials, equipment, production, and distribution. The selection of the proper site and facility can make a valuable contribution to a company's profitability.

Real estate costs are a business expense, and as such do not produce a profit. Therefore it is vital that you carefully analyze your real estate needs to determine just what you do and do not need.

After you have determined that your present site and/or facility is inadequate for present or projected use, and your company decides to move, you must be prepared for a site search.

Before any site search work begins, you must make one major decision. Do you want to conduct the site search in secrecy, or not? That option is yours, but consider the ramifications before you make the decision.

Some companies *announce to the world* that they are seeking a new site. For some this is good publicity, their product or service may be significantly enhanced by the announcement of growth or expansion. For others it could be very destructive. Only you know what you must do.

The most serious problem occurs when the plan is to relocate from one area to another. Many employees look upon this as abandonment, fearing the loss of their jobs. Union leaders may even attempt to block this move, and rightly so. Frequently, local politicians may become involved. This has occurred many times where companies from the northeast planned to move to the sunbelt region.

Land cost inflation can also sometimes be a problem. This is not uncommon when a large and prestigious company is seeking land in a community. Land prices can escalate significantly overnight. However, when a property is sought in adjacent communities or states, the bidding for this new industry can become competitive.

There is no common answer for every company in every situation. However, the following suggestions may be helpful: Premature disclosure of your intention to move should be avoided. Initial investigation should be done in complete secrecy, and we suggest you use discretion in the early phase of site search.

REASONS FOR SITE SEARCH

There are a number of reasons for a site search. Here are a few:

1. The present site is inadequate.
2. The company is expanding into a new market area.
3. Operations are being decentralized.
4. The company is relocating to an area adjacent to an existing or new source of raw materials.
5. The company is relocating to a region where distribution of the finished product can be centralized.
6. Operations are being consolidated at a new location.
7. The company is relocating to an area where specialized and skilled manpower is available.
8. Economics.

USE OF SITE

The type and degree of your involvement in the site search process depends to a large extent on the end use of the site. This could range all the way from a lease of space in an existing building to the purchase of land with a view to major construction for long range corporate development. The following covers site search considerations:

1. Lease space in an existing building at a new location.
2. Lease or purchase land with an existing building.
3. Lease in an industrial/commercial park where the developer will construct buldings to corporate specifications.
4. Purchase land with the buildings to be constructed by a local design/build firm to corporate specifications.
5. Purchase land on which buildings will be constructed to corporate specifications and designed by an architectural/engineering firm.

Long before the site search can or should commence, the need for the site must be established. It is then necessary to define the requirements, which will vary to some extent, depending upon the specific use of the site. The type of site required is different for an industrial, commercial, institutional, or academic facility.

All of this may lead you to believe that the task of site search and selection is horrendous and imponderable, but it needn't be so, providing you approach it in a systematic fashion, and you follow a simple step-by-step procedure.

Since this chapter deals with site search and selection, we will proceed on the basis that you have determined *why* you need the new site; we will address the *how* of site search and selection.

PHYSICAL FACILITY REQUIREMENTS

The first step is to determine the physical requirements of the new facility. This is best accomplished by developing a simple checklist noting all the items to be considered. This checklist must be developed for the specific facility, for as explained earlier, the needs of an industrial plant are vastly different from those of a hospital.

The following list was developed for a manufacturing plant with an adjacent office building. It was not intended to cover every need of every type of manufacturing operation. It is included here to show what items should be considered. You may have additional specific requirements, in that case, just add them to the list. This list will then be tailored to your individual needs.

1. Space in square feet
   a. Plant __________
   b. Office __________
2. Height in stories
   a. Plant __________
   b. Office __________
3. Clear height: plant __________
4. Column spacing: plant __________
5. Floors: type and finish __________
6. Walls: type of construction __________
7. Elevators: size, type, and capacity
   a. Passenger __________
   b. Freight __________
8. Utilities and fuel __________
   a. Natural gas __________
   b. Steam __________
   c. Water __________
   d. Sewer __________
   e. Oil __________
   f. Coal __________
9. Electric power
   a. Voltage: primary __________
   b. Voltage: distribution __________
   c. Load, kva __________
   d. Phase __________
   e. Frequency __________

10. Lighting
    a. Fixture type ______
    b. Foot candles ______
11. Heating, ventilation, and air conditioning
    a. Plant ______
    b. Office ______
    c. Ventilation ______
    d. Exhaust ______
12. Number of employees
    a. Male: plant ______
    b. Female: plant ______
    c. Male: office ______
    d. Female: office ______
13. Toilet facilities
    a. Plant ______
    b. Office ______
14. Lockers ______
15. Showers ______
16. Drinking fountains ______
17. Emergency power and lighting
    a. Plant ______
    b. Office ______
18. Fire protection
    a. Sprinklers ______
    b. Halon system ______
    c. Fire extinguishers ______
    d. Smoke detection ______
    e. Smoke evacuation ______
19. Security
    a. Perimeter surveillance ______
    b. Fence alarm ______
    c. Door alarms ______
    d. Closed circuit TV ______
    e. Entrance and exit monitoring ______
    f. Central control and monitoring ______
20. Cafeteria
    a. Kitchen ______
    b. Vending machines ______
    c. Seating capacity ______
    d. Operating hours ______
21. Computer room ______
22. Medical facility ______
23. Parking
    a. Employee: car spaces ______
    b. Visitor: car spaces ______
    c. Lot size: area ______

24. Truck docks, number ________________
25. Railroad docks, number ________________
26. Ship dock facility ________________
27. On-site storage—raw materials ________________
28. On-site storage—finished goods ________________
29. On-site storage—liquid and solid fuels ________________
30. Bulk conveyor system ________________
31. Liquid pipe lines ________________
32. Flammables storage ________________
33. Explosives storage ________________
34. Total site size, acres ________________

Many of the items on this list require nothing more than to be included so that they can be checked off during the initial phase of planning for the new facility. Some will require serious consideration depending upon their value to the new facility. Others will be investigated by the design professional, and whether required by code or because they represent good engineering practice, he will determine the quantity or size required.

This first step, while actually not part of site search and selection, was included to illustrate what preparation is necessary before site search can be undertaken. You need to know the size and type of building, plant, and grounds you require so that you can best match the site to your facility.

PHYSICAL SITE REQUIREMENTS

The next step is to determine the physical requirements of the new site. These requirements should also be noted in the form of a checklist in order to insure that all essential items are considered and evaluated. With a properly prepared checklist it is a simple matter to provide uniform and meaningful evaluation of all sites involved.

These are the items that will permit a functional use of the facility in order to accomplish the intended objective. This, as you can see, includes more than just the physical aspects of the bounded site, here you must also evaluate access to the site, on-site, and off-site transportation, and provisions must be made for bulk materials, fuels, finished product, etc., storage, where appropriate. No essential item will be missed by anyone who has your checklist. We strongly recommend that you develop your own list very carefully, noting all necessary items you need considered.

1. Size of property, acres ________________
2. Transportation services
    a. Waterways ________________
    b. Railroads ________________
    c. Motor trucking ________________
    d. Public transportation ________________
    e. Airport ________________

3. Utilities
   a. Electricity
      1. Voltage: primary ________________
      2. System capacity, kva ________________
      3. Phase ________________
      4. Frequency ________________
   b. Natural gas ________________
   c. Steam ________________
4. City water
   a. For domestic use ________________
   b. For fire protection ________________
      i. Pressure, psi
      ii. At flowrate, gpm ________________
5. Public sewerage ________________

REGIONAL CONSIDERATIONS

You are now at the first major crossroad of your site search and selection program. You must now determine in what region of the country you plan to locate the facility. You should consider the reason for the site, the end use of the site, and if you can, determine the preferred location of your chief executive.

We were personally involved two decades ago in site search for the division headquarters and primary manufacturing facility of a major corporation. The number one criterion, developed by the executive, was suitable labor supply—in this case, electronic assemblers, technicians, and engineers were required. After several unsuccessful proposals, we were subtly informed that the chief executive was born and raised near one of the sites and he had a summer retreat in the mountains within 50 miles. That concluded our site search for a specific region. We tailored our next proposal appropriately and we received approval to proceed immediately. This short case history is included to show that sometimes site search begins and ends in the ivory tower.

Chief executives notwithstanding, regional considerations must be addressed. While this point cannot be covered as simply as the others by a checklist format, we will list the various items, appropriately grouped, with a short discussion as necessary.

1. Climate. In many instances, this will be of casual importance and it will not be a significant factor in the final regional selection criteria. In other cases, for example, poultry and livestock processing, agriculture, etc., it may be necessary to determine the tolerable limits of the following:

a. Temperature average: annual ________________
b. Temperature average: seasonal ________________
c. Temperature maximum: high ________________
d. Temperature minimum: low ________________
e. Average degree days: winter ________________
f. Average degree days: summer ________________

g. Rainfall average: annual __________
h. Rainfall average: seasonal __________
i. Snowfall average: seasonal __________
j. Humidity average: relative __________

2. Transportation. The type, availability, and rates of transportation require serious consideration, particularly in those cases where raw material and finished product must be transported into and out of the plant. For example, a sugar-processing plant will require adequate waterways, railroad connection, and reasonable accessibility to highways, whereas an office building will require, at most, some means of public transportation for employees.

Provisions for off-site transportation must be included in the planning stage, i.e., provide connections and access to railways, highways, waterways, pipelines, and airfields for transportation of raw materials, supplies, liquid and solid fuels, parts, finished goods, and personnel.

On-site transportation requires equal consideration. This could include roadways, railroad sidings, water frontage, access to storage areas, and stockpile sites. On-site transportation should include personnel transport, automobile parking and driveways, heliport and landing strips, and bus loading and unloading area.

This list covers freight and passenger transportation:

a. Type of waterways __________
b. Length of waterway frontage __________
c. Description of system __________
d. Waterway freight rates __________
e. Distance to major highway __________
f. Available trucking service __________
g. Restriction on trucking service __________
h. Motor freight rates __________
i. Distance to major railroad __________
j. Available trunk lines __________
k. Restriction on railroad service __________
l. Railroad freight rates __________
m. Distance to nearest airport __________
n. Available freight service __________
o. Commuter railroad service __________
p. Commuter bus service __________
q. Quality of taxi service __________
r. On-site roadways __________
s. Railroad siding __________
t. Access to storage area __________
u. Automobile parking area __________
v. Driveways __________
w. Heliport and landing strip __________
x. Bus loading and unloading area __________

3. Taxes. Frequently taxes could make the difference between profit and loss for the operation. Many communities offer tax incentives for new construction, and such proposals require appropriate evaluation. Unemployment compensation and payroll taxes vary widely across the country and should not be overlooked. Compare and evaluate these taxes:

a. Real estate ____________
b. Personal ____________
c. Sales ____________
d. Corporate income ____________
e. Corporate payroll ____________
f. Personal income ____________
g. Incentives ____________

4. Financing. Many communities, cities, and states offer as inducements low-interest, long-term financing for new plants and plant expansions. These should be compared and evaluated for the various site locations before final selection is made.

5. Labor. While many facilities managers carefully check and thoroughly evaluate prevailing labor rates, other considerations are frequently overlooked or casually dismissed because the site location region seems right for the particular personnel needs of the plant or office. It is important that suitably skilled labor be available in sufficient quantity to staff the new facility or replace those people who elect not to relocate. Compare and evaluate the following:

a. Prevailing labor rates ____________
b. Availability of suitable labor ____________
c. Unemployment rate ____________
d. History of labor disturbances ____________
e. Total available labor force ____________
f. Productivity of suitable labor ____________
g. Funded employee training ____________
h. Vocational/technical schools ____________
i. Right-to-work laws ____________
j. Social composition of population ____________
k. Median income of population ____________
l. Industrial/commercial mix ____________

6. Miscellaneous considerations. This is by no means less important than the above, its value depends entirely upon the company's specific needs. These items are provided for your evaluation, when considered necessary:

a. Education. Type, quantity, and quality of all levels of education available including occupational/vocational training.

b. Recreation. This must be considered and should be consistant with the needs of your employees. This is most important when considering relocating employees from one region to another, as they are accustomed to a type and quality of recreation that may not be available at the new location. (Local labor is usually satisfied with available local recreation.)

c. Housing. Is there sufficient housing of the type and quality required by the relocated employees?

d. Fire and Police protection. Consider the type, quality, performance, and responsiveness of the local fire and police departments.

e. Community Services. Are there adequate medical care facilities? Is the road and highway maintenance satisfactory? Are the sewer and water services adequate to accommodate your requirements without major additional expense?

ZEROING IN ON A SPECIFIC SITE

Now that you have completed your site selection process and have determined the appropriate region and area for your new facility, you must zero in on a specific site within the general area that is most compatible with your particular facility and that best fits your requirements. Actually, this phase of site selection is very possibly the most difficult, the least understood, and frequently, it is simply brushed off.

It is not unusual for a company to spend a great amount of time and money on the site selection process and then pick any piece of ground, sometimes the cheapest available, in the general location, and then have its architectural/engineering (A/E) firm tailor the design of the facility to the site.

Since not all sites are flat and not all buildings are one story, the compatibility of the site and building is an essential consideration, prior to the purchase of the tract of land. And the passing of each day, new and more complex guidelines, regulations, and codes impact on the siting of the building. In recent years, with the advent of the energy crisis, even the orientation of a building requires careful planning. The environmental impact can be very important, for example, flood-proofing regulations, soil erosion and sedimentation control, clean-stream laws, to mention a few. There are zoning ordinances and, in addition to local and state laws, of course, there are the Environmental Protection Agency regulations.

As you can anticipate from the above, site selection in a general region completes only part of the project. This portion delineates the best use of real property resources to serve the assigned and projected mission. The arrangement of facilities should be integrated with the site to achieve optimum land use, good functional relationship, environmental harmony, and a pleasant appearance consistent with short- and long-term economical objectives.

We will now describe broad technical criteria and provide guidance for design and construction of functional, durable facilities which have reasonable and appropriate maintenance and operating costs over their design life. Appropriate consulting services should be brought into the picture at the earliest date. Their services will cost as little or as much as you want to spend.

1. Siting of buildings. Buildings should be located in a functional relationship to each other for operational efficiency and economical maintenance, and in consonance with the topography for construction, economy, and preservation of the

character of the site. Setbacks and spacing of buildings must provide for the admission of adequate light, circulation of air, fire safety clearances, pedestrian access, vehicle access, and parking where required.

The composition and relationship of structures, roads, and utilities are of equal importance to the individual design of these features in producing arrangements that are efficient to operate and maintain, economically acceptable to the topography, and conducive to a pleasant working environment to avoid monotony and regimentation.

For a building to be economically acceptable to the topography, the amount of cut and fill required must be given careful consideration as this is an expense that provides no return on investment. Further, the construction of a building on new fill can create insurmountable problems in the future because of settlement, or uneven settlement, manifesting itself in cracking of walls and floors. Even roof and wall separation has been known to occur. Any fill required must be placed in accordance with accepted engineering practices, using only quality material satisfactory for the purpose.

Soil and foundation conditions must be investigated to assure suitability for economical excavation, building foundations, site preparation, and grading. On sites where the soil is of questionable quality, money spent on test borings and soils analysis is insurance against future problems.

Buildings should be oriented to take advantage of views, topography, trees, favorable weather exposure, and other site features to the extent that such orientation is feasible. Building orientation is of particular importance today. The first cost of HVAC equipment must be considered, as well as subsequent cost of operation of HVAC systems and equipment.

For air-conditioned buildings, it is important that a minimum area of the structure be exposed to the afternoon sun and that there be only minimum solar load on fenestration morning and afternoon. Proper orientation in relation to prevailing summer breezes is especially important for buildings that do not have air conditioning or mechanical ventilation.

With the advent of solar heating, it is now necessary to plan for the installation of solar panels. Frequently these are located on the roof of the building, but do not overlook other options. One company, fortunate to have an employee walkway from the plant to the parking lot of appropriate orientation, placed the solar panels over this walkway. In addition to an aesthetically integrated installation, the employees obtained a weathertight, covered walkway.

These panels have also been installed on ground level at some distance from the plant where they can be integrated into the site and to some extent hidden from view.

As you can see, there are several options for the location and installation of solar panels. But keep in mind that these solar panels must be so placed that they receive direct sunlight and that they have correct orientation to the sun.

Since there are no laws at this time that govern the right-to-sunlight, be aware that someone, someday, may build in such a manner that your solar panels may be obscured from direct sunlight.

2. Grading, Drainage, and Erosion Control. Grading, drainage, and erosion control are definitely interelated. What is done in one area could greatly affect the others. Frequently, the best solution for these items is that nothing be done. Important consideration must be given to the preservation of the natural character of the terrain by

minimum disturbance of existing ground forms and vegetation cover. This means minimum earth moving with reasonable balance of cut and fill. Other areas that require consideration are: avoidance of earth banks that entail costly measures for erosion control; use of a minimum number of steps to entrances; creation of satisfactory grade levels around buildings and preservation of existing trees.

In order to provide positive drainage, a slope of not less than two percent for turfed areas is desirable, however, more permeable soils, such as sandy types, may need a lesser slope. Drainage slopes away from buildings should be a minimum of six inches vertical for each horizontal distance of ten feet. Banks for transition from one area to another should not be steeper than one foot vertical in three feet horizontal. Steeper slopes may be used with rock outcrops or vegetation cover, other than turf, which can be easily established and maintained.

Control measures for erosion of surface soil to rehabilitate areas scarred or denuded during construction may include terraces, ditch drop structures, course-graded aggregate, anchored mulch, sprinkler irrigation systems in low rainfall areas, and vegetation. Let it suffice to state that, insofar as practicable, uneconomic, hazardous, and unnecessary use of flood plains and areas subject to flash flooding must be avoided.

3. Vehicular/Pedestrian Circulation. The best system of vehicular and pedestrian circulation is one that, as far as practicable, isolates one from the other. Pedestrian access to buildings should never be directed across vehicular-access roads. Through traffic for the facility should be kept to a minimum.

Convenient and safe vehicular access and circulation should be provided for essential services, such as deliveries, trash and garbage collection, fire protection, emergency vehicles, maintenance and repair vehicles, and should be as remote as practicable from pedestrian traffic.

Provide off-street parking, adjacent to but unobtrusively related to the facility served. Sidewalks should be designed to provide convenient and safe pedestrian access and necessary circulation. Width of the walks depends upon the anticipated pedestrian traffic volume. Grades of the walks should follow the natural pitch of the ground, avoiding steps, particularly single risers, because in many cases access for the handicapped must be provided.

4. Landscaping. Appropriate landscaping is essential. It can be used to hide or disguise ugly elements of the facility and to enhance its natural beauty. Primary objectives of landscape plantings are to:

a. Enhance the exteriors
b. Integrate the buildings with the adjacent open areas
c. Establish and define land-use areas
d. Provide appropriate scale and character to buildings
e. Provide necessary privacy
f. Reduce air-conditioning requirements by providing shade in summer
g. Reduce heating requirements in winter by providing wind and snow breaks
h. Screen out views and hide less desirable features
i. Reduce local noise
j. Provide dust and erosion control

5. Architectural Considerations. The whole of architecture, as practiced by profit-making firms for profit-oriented companies, is well summarized in the following one-sentence statement: *Excellence of architectural design should rank with economy of construction and functional efficiency as prime requisite for all construction projects.*

The statement is rather simple, but the implementation is difficult. The trick is to have excellent design with economy of construction for functional efficiency.

In the not too distant past, and even to some extent today, particularly for the construction of buildings for resale, i.e., spec buildings, initial investment was and is the prime consideration. Now, life-cycle costing is getting the appropriate attention it deserves, and rightly so. Life-cycle costing takes into account initial cost, energy cost, operation and maintenance cost, interest rate, and energy escalation rate. With life-cycle costing you obtain the true cost of your investment for comparison with alternate systems, equipment, and materials.

Building configuration and envelope, as well as orientation, determine the amount of energy consumed and the initial investment. For example, a circular building has less exterior surface, hence would cost less to build, and has less heat gain or loss than any other shape for equal floor space. A square building has less surface than a rectangular building, therefore it would cost less to build and would have less heat gain or loss.

The building envelope, consisting of solid walls, windows, doors and other openings, roof and floor surfaces, is affected by the climate because of the envelope's orientation.

The total cost of the building should be considered when selecting a structural system. Also, the architectural design, because of its impact on the structural system, should be developed with appropriate consideration for the structural system. The cheapest structural system does not necessarily yield the lowest total building cost. Keep in mind that the structural system affects the floor-to-floor height, and of course, the higher the building the more it will cost to construct. In trying to keep the floor-to-floor height as small as possible, remember that the ceiling plenum, the space above the ceiling and below the structural members, must be adequate for the myriad of pipes, ducts, etc., that are installed in this space. All too frequently this space is too small.

When selecting structural materials and systems for a specific project, consider the following items:

1. Site environment
2. Subsurface conditions and frost penetrations
3. Wind velocity and prevailing direction
4. Seismic rating
5. Snow loads
6. Typhoons, hurricanes, and tornadoes

The architectural features of the building must be compatible, externally, with the environment, and must permit integration of the electrical and mechanical systems, internally, into the building. There was a time, and isolated cases exist even today, when architects completed the design of a building, then expected the engineer to put in the electrical and mechanical systems. To make matters worse, even electrical and mechanical engineers often performed their designs independently. Fortunately, this has changed over the years, and many A/E firms now establish project teams. With this scheme, a group of architects and engineers work together to design a building.

To cite an example of changing architectural concepts, people once lived in caves. With the passing of time, they emerged to live and work in monstrous edifices built as monuments to the designer. Recently, particularly with emphasis on energy conservation, people are moving back into caves, as buildings are being built into the ground, underground, and into hillsides, not necessarily because architecture alone dictates the design, but also in deference to energy conservation.

In the past decade there have been numerous excellent, functional architectural design concepts where uneven terrain was used for multilevel structures to fully utilize building orientation for effective energy conservation. In many of these cases, the direction of the summer sun and the prevailing winter wind was given appropriate consideration in the orientation of the structure long before the design of the building was started.

The most important aspect of zeroing in on a specific site is to make optimum use of real property resources to satisfy the objective and to try to fully utilize, to best advantage, the natural features of the land. A building appropriately designed and properly integrated with the site will appear to be part of the site. Zeroing in is well worth the effort.

SUMMARY

Here are several important considerations, do's and don'ts, when zeroing in on a specific site.

Some do's:

1. Do include the A/E firm early in the planning stage.
2. Do select building orientation carefully.
3. Do give consideration to site terrain.
4. Do retain trees for wind breaks and for aesthetic reasons.
5. Do consider site climatic conditions.
6. Do select a building envelope compatible with the environment.
7. Do integrate electrical, mechanical, and structural systems with the architectural design.

Some don'ts:

1. Don't force the building onto the site.
2. Don't denude the site of trees.
3. Don't bulldoze the terrain flat.
4. Don't cross pedestrian and vehicular traffic.
5. Don't place noisy equipment or operation next to quiet areas.
6. Don't purchase a site where adjacent buildings or operations are not compatible with the planned facility.

2. Leasing Considerations

Many corporations are well aware of the importance of real estate leasing, so much so that frequently vice-presidents are assigned the responsibility of heading-up the real estate department. Some corporations utilize an independent division or company, headed by a president, to handle this function. For every major corporation that is adequately staffed with qualified and experienced professionals, there are thousands of smaller companies that lack this expertise. But these people can execute successful leasing arrangements if they follow a simple step-by-step procedure that their experienced counterparts use.

Most difficulty is experienced by those who are required to lease space infrequently, or are involved in the process for the first time. One of these days, you may be called upon to provide plant or office space on short notice; for temporary or permanent purposes; or for specific needs. Regardless of how your needs are best served, you should use a systematic approach to the problem. There are certain basic steps to follow in order to insure a successful execution of a lease.

1. Determine the type of lease needed.
2. Determine the amount of space required.
3. In cases where it is anticipated that leasing will be required on a regular and frequent basis, develop a leasing procedure.
4. Conduct a lease investigation and comparison of several sites.
5. Complete lease agreement.

TYPES OF LEASES

The type of lease sought and obtained depends upon the reason for the lease. While this sounds like an obvious statement, it isn't necessarily so. All too frequently companies get locked into leases that do not truly reflect their needs. Leases could run the gamut from a month-to-month type to a multiyear type with numerous options. There are also emergency situations that require expediency, and there are landlords that will provide leases to suit any situation. In these cases, it's just a simple matter of matching your needs with a landlord's space and availability.

The simplest solution to the problem of providing the needed space is to contact a realtor, property manager, or whatever he is called, sign an agreement and move in. If you are lucky, you may have made a sound decision, but luck is not what you need. You need a practical approach to building leasing. You need to know your requirements, then locate the building that most nearly satisfies your needs at an optimum cost.

Regardless of the type of lease involved, you must first determine the amount and type of space needed. In those cases where a long-term lease is considered, you will also need to determine the amount of space required at different periods of the lease. The lease must be written to accommodate these anticipated changes in space. If growth is not projected, the lease agreement becomes simpler. When expansion space is considered, options must be included in the lease agreement. More about this later.

ORGANIZATION CHART

Office space planning can be a rather simple task, provided that a basic procedure is followed. If you will complete the tasks outlined, you will be able to determine the amount and type of space required, at the time of the survey and at any future date, as programmed.

1. Develop the organization chart
2. Establish space and furniture standards
3. Complete the projected space requirements

Figure 2–1, An Organization Sketch Sheet, provides a simplified guide for determining the organization (company, division, department, etc.) reporting levels that are essential for tabulating the size of space assignable to the various positions. When properly completed, the sketch sheet shows: reporting levels, areas of responsibility, departmental subdivisions, and total number of employees.

It is important that this chart be properly constructed because it will be used to determine the size of space occupied and the facilitation to be provided. For example, the president of a small company may have one vice-president, two managers, and three supervisors reporting directly to him. These people, who occupy different levels on the chart, will need different size offices and facilitation.

Fill out the form by placing the name of the highest level position in the lone block at the top of the chart. Following the example in Figure 2–1, the vice-president is entered in the top row, the managers in the following row, and the supervisors in the third row. Clerical (or other groups) are noted in the bottom row by title and number.

The organization sketch sheet is also a valuable management tool that can be used for everyday operation. The block in the upper left-hand corner is used to identify the organization, its location, etc. The sketch sheet, when completed, constitutes the organization chart. Depending upon the size of the company, one or more sketch sheets may be required. When the appropriate number of sheets are completed, you are ready to move on to the next step in space planning.

SPACE AND FURNITURE STANDARDS

The advantages of standardization in industry are well known and universally recognized. Utilizing standards promotes the adoption and use of the best planning methods, it facilitates space and furniture planning, and it saves time and money. The

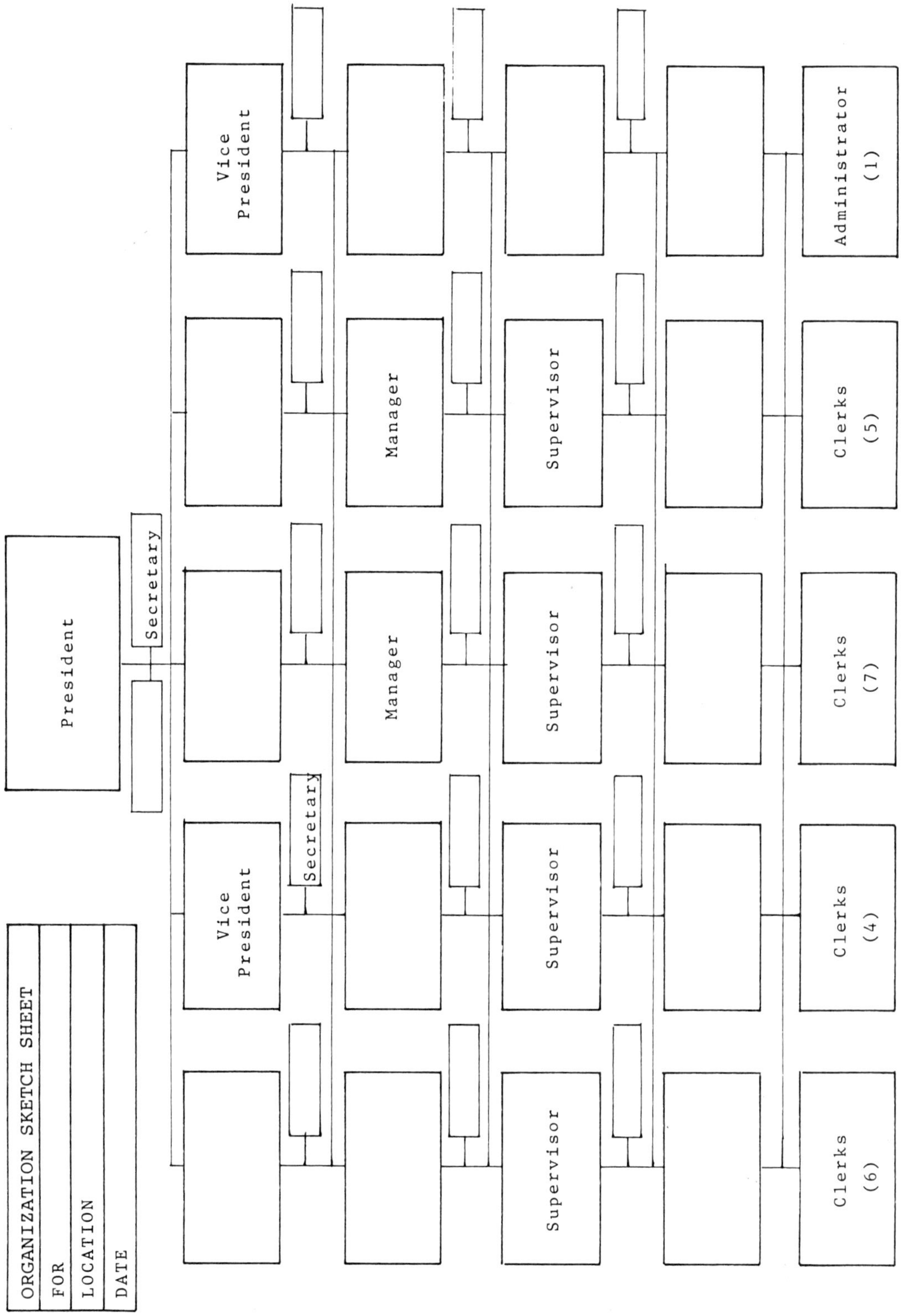

Figure 2-1. Organization Sketch Sheet.

Table 2-1. Office Space Standards.

SPACE GROUP	OCCUPANT OR USE	AREA, SQ FT	TYPE OF SPACE
SP	President	400	Enclosed,
A	Vice-president	300–320	Ceiling High
B	Department Manager	225–256	Partitions
C	Section Manager	120–150	
D	Supervisor	100–120	Cubicle
E	Clerk	54	Open
F	Clerk	48	Floor
	Secretary	50–120	Varies
CONFERENCE ROOMS			
R	Conference for 18	450–512	Enclosed,
S	Conference for 14	375–384	Ceiling High
T	Conference for 10	225–256	Partitions
U	Conference for 8	150–192	
V	Conference for 6	144–150	

purpose of space and furniture standards is to establish a consistent program that can be utilized throughout the corporation for the assignment of uniform office and space areas, and for the selection and implementation of functional furniture, adequate to the task. It also serves as a guide for space planning, cost budgeting, and facilities leasing. A properly developed space standard is designed to provide an adequate area for the furniture intended for the position level involved.

The space and furniture standard presented here can be used as a guide for developing your own specific standards. The guide standard has three office types: ceiling high for executive offices and conference rooms; half high or bank screen partitioned cubicles for supervisor offices; and open floor space for clerical workstations.

Please note that the spaces assigned to various groups contain areas of two different dimensions. This is necessary since most buildings are constructed on either the four-foot or five-foot ceiling module, which will then permit all partitions to match the ceiling grid.

The space group designation for the position of president is SP, indicating "special," and the size has been set at 400 square feet. That area should be based on the individual's specific method of operation.

Finally, there is nothing magic about the sizes indicated for the various groups. When you develop your standard, you may want to tailor it to your specific needs. Keep in mind that these group types and sizes are the generally accepted standards for industry.

DETERMINING SPACE REQUIREMENTS

Figure 2–7, the Projected Space Requirement Form, is a simplified guide for determining current and projected personnel growth and developing current and future

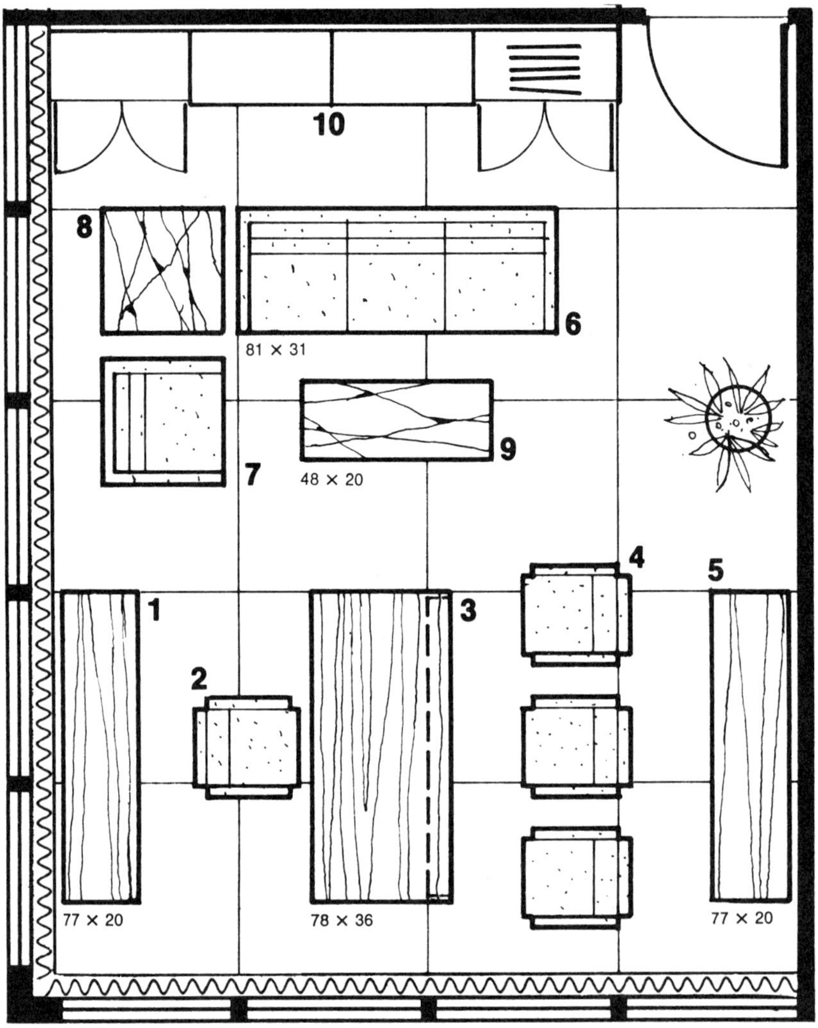

Figure 2-2. Office Space Group A complete with furniture. See Figure 2-3 for furniture identification.

space needs. It was designed for a two-year projection, but it can be modified to accommodate any number of years to satisfy individual needs.

Actually, columns 2 through 5 and items 9 through 14 should be filled in by the particular department or group involved, as they know the number of personnel currently under their control and are in the best position to determine future growth.

Column 2: Enter the various titles and job classification for all personnel.
Column 3: Enter the number of personnel presently functioning in each of the positions indicated in column 2.
Column 4: Estimate and enter the total number of personnel to be functioning in each of the positions indicated in column 2 in 1 year from the present date.
Column 5: Estimate and enter the total number of personnel to be functioning in each of the positions indicated in column 2 in 2 years from the present date.

Figure 2-3. Furniture Group A for Space Group A.

Item 9: Indicate conference rooms required by the number of rooms and the number of people using each room.
Item 10: Indicate the types and numbers of files.
Item 11: Indicate the types and size of coat storage.
Item 12: Indicate any need for general storage space.
Item 13: Indicate the number of daily visitors anticipated.
Item 14: Indicate miscellaneous and unique requirements not covered under any of the above categories.

Using this input, you can now complete the form to determine current and projected space requirements. This now becomes a very simple matter of plugging in numbers you take from your office space standards.

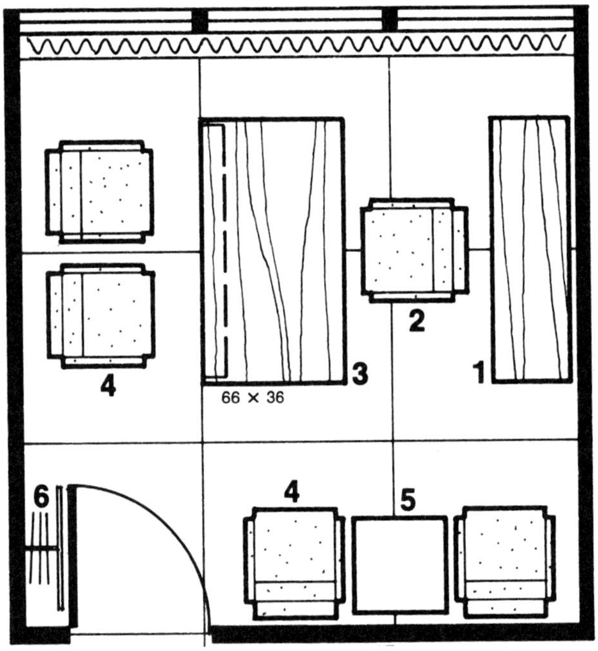

Figure 2-4. Office Space Group C.

Column 1: Enter space group designation of each person indicated as determined from the space standards.

Columns 6, 7, and 8: Enter space requirements based upon the various space groups and the number of personnel in each.

Items 9 through 17: Enter space requirement for the functions indicated and the total space required.

When the form is properly completed, the bottom line (usable total) will show the amount of space in square feet that is currently needed and the amount projected into some future period of time. Usable total square feet does not include stairways, elevator shafts, toilet rooms, mechanical rooms, and other space that cannot be occupied.

Depending upon the size of the organization, the appropriate number of projected space requirement forms must be completed. The sum of all the spaces establishes the amount of usable space required by the lease. In cases where significant growth is projected, options must be included in the lease to accommodate the expansion required. It is normally not economical to lease more space than will be filled within a year or two. Good judgment will be the determining factor when selecting the projected growth period to be included in current leasing considerations.

You should by now have determined the amount of space you will need, currently, and in the future. The next logical step is to seek and obtain a lease. This is best accomplished through the use of a standard procedure to simplify the project, insure uniformity, maintain control, and to provide results within a reasonable time frame and cost.

Whether the need is for an occasional real estate lease, or because leasing is on a

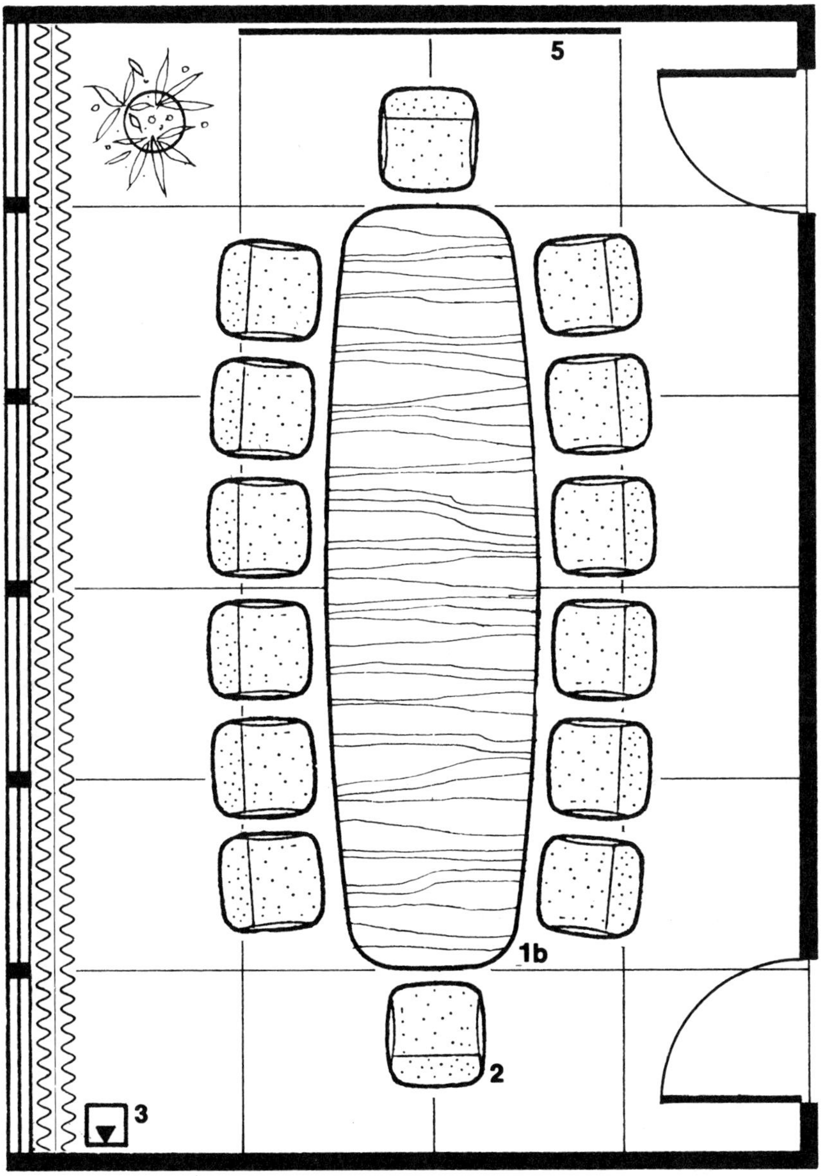

Figure 2-5. Conference Room Space Group S.

regular basis, the procedure outlined below can serve as a valuable management tool. The real estate leasing procedure defines the responsibilities of the various activities of the company and establishes the time frame within which these responsibilities must be accomplished. It establishes the authority of the department responsible for real estate leasing, and excludes other activities from "committing" leasing of real estate or purchasing leasehold improvements. It defines the various situations under which the procedure exhibits control. This procedure provides the means for uniform and consistent site search and evaluation, and is an effective tool for control of real estate leasing.

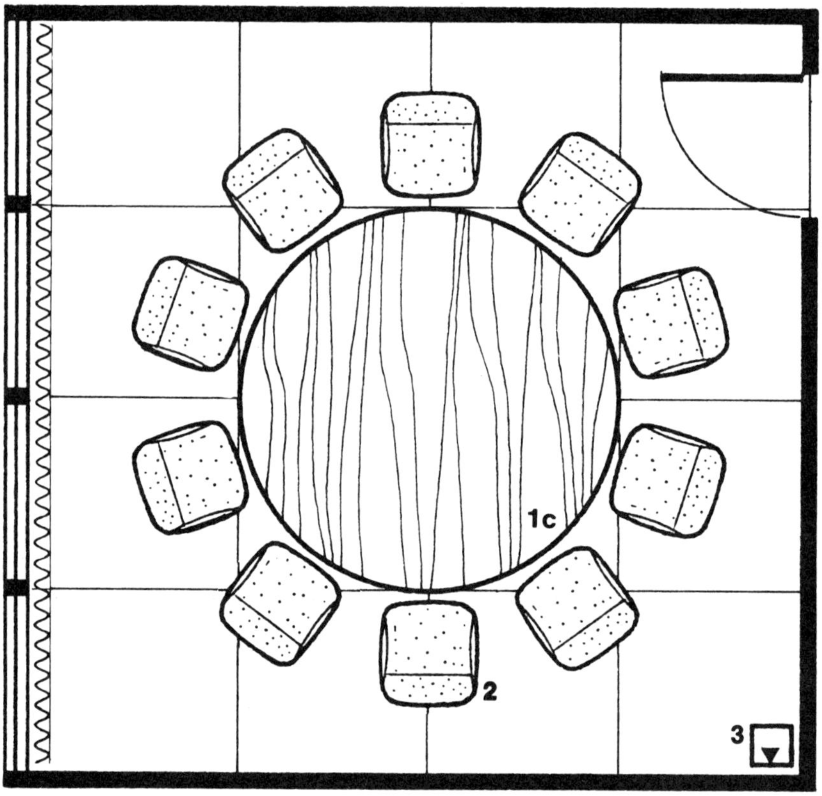

Figure 2-6. Conference Room Space Group T.

REAL ESTATE LEASING PROCEDURE

The following procedure is the result of numerous refinements over the years and has been utilized successfully in both a large corporation and a small company. In the event that your requirements differ from those outlined in the procedure, you can tailor it to your specific needs.

Real Estate Leasing Procedure

I. Introduction

The purpose of this document is to establish the procedure to be followed for the processing of requests for leased real estate and/or leasehold improvements. Any and all commitments for the leasing of real estate and/or the purchasing of, or contracting for, leasehold improvements shall be made only by the Facilities Department after proper authorizations have been obtained. *Under no circumstances shall any other activity, verbally or in writing, commit for the leasing of real estate or for the purchasing of, or contracting for, leasehold improvements, or make any statements, oral or written, that could be construed to legally bind any member of "Abc Inc." in such matters.*

Scope

All companies within the Abc Group are governed by this procedure.

PROJECTED SPACE REQUIREMENTS FORM

Department Date

Space Group	Title	Personnel Projection Present	+ 1 Year	+ 2 Year	Space Projection Present	+ 1 Year	+ 2 Year
(1)	(2)	(3)	(4)	(5)	(6)	(7)	(8)
SP	President	1	1	1	400	400	400
A	Vice President	1	1	2	300	300	600
B	Manager, Dept.	2	3	4	450	675	900
C	Manager, Sect.	4	5	8	600	750	1200
F	Clerks	28	40	60	1400	2000	3000
-	Secretarial	2	3	7	200	270	550
-	- - -	-	-	-	-	-	-
-	- - -	-	-	-	-	-	-
-	- - -	-	-	-	-	-	-
Conference Room	(9)	1 - S - 14 People			375	375	375
General Filing	(10)	25 Files - Letter Size			200	300	400
Coat Storage	(11)				50	100	100
Lobby	(12)	15 People			300	300	300
Miscellany	(13)	Library			150	150	150
Storage Space	(14)	Paper Stock			196	196	196
- - -	- - -				-	-	-
Sub-Total	(15)				4771	5966	8321
Traffic Area	(15 to 20% of above sub-total) (16)				954	1193	1664
Usable Total	(17)				5725	7159	9985

Figure 2-7. Projected Space Requirements Form.

Table of Contents

- I. Introduction
- II. Procedure
  - A. Existing Leases
    1. Renewal
    2. Termination
    3. Relocation
    4. Expansion

B. New Leases and/or Leasehold Improvements
1. All new leases and/or leasehold improvements
2. Space less than 1,500 square feet
3. Space greater than 1,500 square feet
C. Space Requirements
D. Management Approval

II. Procedure

A. *Existing Leases*

This procedure is to be followed, for the renewal of real estate leases that do not involve the purchase of leasehold improvements, for the termination of real estate leases, and for the relocation or expansion of existing facilities.

The Facilities Department will notify the occupying activity of the lease expiration date sixty (60) days in advance of the required landlord notification date.

1. *Renewal*

a. The occupying activity will forward to the Facilities Department through their company president a letter of recommendation for renewal of the existing lease thirty (30) days prior to the required landlord notification due date.

b. The Facilities Department will then obtain the necessary management approval and process the necessary documentation to renew the lease.

2. *Termination*

a. The occupying activity will forward a letter recommending the termination of the existing lease to the Facilities Department through their company president thirty (30) days prior to the required landlord notification due date.

b. The Facilities Department will then obtain management approval and process all necessary documentation to terminate the lease, and provide for restoration and all other termination provisions of the lease as applicable.

3. *Relocation*

a. The occupying activity will forward a letter recommending the relocation of the existing facility to the Facilities Department through their company president thirty (30) days prior to the required landlord notification due date.

b. The Facilities Department will then obtain management approval to relocate the facility and terminate the existing lease, according to Section II-A, 2-b above. The procedure then reverts to Section II-B for establishing the new lease.

4. *Expansion*

a. The procedure to be followed for the expansion of existing facilities is that as prescribed in Section II-B for establishing new leases and/or leasehold improvements.

b. The Facilities Department will then obtain the necessary management approval and process the necessary documentation to renew the lease.

B. *New Leases and/or Leasehold Improvements*

This procedure is to be followed for the processing of requests to lease new real estate and/or the processing of requests for leasehold improvements.

1. *All New Leases and/or Leasehold Improvements*

a. The operating activity will forward a letter of justification detailing the need for the acquisition of leased real estate and/or leasehold improvements based upon personnel and business requirements. For minor leasehold improvements, a sketch of the proposed alteration should be included with the letter of justification.

b. The Facilities Department will review the request and forward to the operating activity recommendations for the amount of space to be leased and the basic terms of the lease. In cases of alterations, the Facilities Department will review the request and prepare working drawings and specifications for the leasehold improvements if deemed necessary. If the recommended space is less than 1500 square feet and/or the leasehold improvements are to be made at the expense of others, the procedure, as indicated under (2) below, will apply. In all other cases, the procedure indicated under iii below will apply.

2. *Space Less than 1,500 Square Feet*

a. For new leases which include Facilities Department recommendations of less than 1500 square feet, and when the operating activity has personnel located within the geographical area, the operating activity will contact local commercial real estate brokers and locate available space that conforms to the guidelines provided by the Facilities Department. A minimum of five (5) such sites should be located. The operating activity will forward this information through their company president to the Facilities Department sixty (60) days prior to the proposed occupancy date.

b. The Facilities Department will then do the following:

i. Prepare a preliminary investigation report for each site.

ii. Prepare a preliminary investigation summary for management review and approval.

iii. Coordinate the final selection of the space to be leased.

iv. Negotiate with the landlord and obtain a lease agreement.

v. Process all documentation.

vi. Maintain liaison with the landlord and the operating activity to coordinate the facilitation of the space. The Facilities Department is the only activity authorized to make changes to the plans and/or schedules for leasehold improvements.

3. *Space Greater than 1500 Square Feet*

a. For new leases which include Facilities Department recommendations for space greater than 1500 square feet, the operating activity will forward through their company president to Facilities Department a report of the functions to be performed at the site, the desired geographical location with reasons for such and all other data which is essential to the geographical location and physical characteristics of the space, six (6) months prior to the proposed occupancy date.

b. The Facilities Department will then do the following:

i. Contact commercial real estate brokers and others, conduct a thorough site survey, and prepare a preliminary investigation report for each site.

ii. Prepare a Preliminary Real Estate Investigation Summary for management review and approval

iii. Coordinate the final selection of the space to be leased.

iv. Negotiate with the landlord and obtain a lease agreement.

v. Process all documentation.

vi. Prepare plans and specifications for leasehold improvements as required. In cases where the leasehold improvements will not be made by the landlord, the Facilities Department will have the total responsibility for the contracting of such work; however, the operating activity may be requested to monitor the work progress because of the geographical location of the space. The Facilities Department is the only activity authorized to make changes to the plans or schedules for the leasehold improvements.

vii. Coordinate the equipment, furniture, and accessories requirement package for the site based upon the Space and Furniture Standards and the unique needs of the operating activity.

viii. Aid the operating activity in the preparation for the physical relocation by assisting in the award of the moving contract, scheduling, etc.

C. In all cases where appropriate (except termination of existing leases) the occupying or operating activity will complete the Projected Space Requirements Form and submit it to the Facilities Department along with other pertinent documentation.

D. Management approval for all leasing and/or real estate activity, shall require the approval of Preliminary Investigation Summary form by the president of the operating activity and the president of "Abc Inc."

PRELIMINARY INVESTIGATION REPORT AND SUMMARY

The next step, of course, is to find real estate that can be leased in accordance with the above procedure. This step is called "site investigation and evaluation." Selecting the proper location for a proposed facility is a very complex operation because there are many items of major concern. Significant savings can be realized or excess money can be better utilized if the project is properly planned. The approach is to utilize a standard system that will simplify the project and insure uniformity in order that sites can be properly evaluated. The backbone of this system are two forms:

- Preliminary Investigation Report, Figure 2-8, 2-9, 2-10, 2-11
- Preliminary Investigation Summary, Figure 2-12

These are used in conjunction with the Real Estate Leasing Procedure. The Preliminary Investigation Report is a 23-item, four-page checklist type of form. It serves a dual purpose: it identifies all the items of concern for proper comparison between the various sites investigated, and it serves as a reminder of all the items that must be questioned in any site search and evaluation. The Preliminary Investigation Report provides the following grouping of information when completed:

- Use of the space
- Purpose
- Location of the site
- Landlord and his agent's address
- Terms of the lease
- Size and cost of the space
- Escalations in lease
- Special clauses in lease
- Services and utilities provided
- Leasehold improvement allowances
- Cost above building allowances

This document is intended for use by the Facilities Department. One set of the forms is to be completed for each site investigated. With this information, the

PRELININARY INVESTIGATION REPORT

PROPOSAL NUMBER____________ DATE____________

1. Unit Requestion Space____________

2. Use of Space____________

____

3. Purpose: New___ Renewal___ Relocation___ Expansion___ Other___
Consolidation___

4. Location: Building Name____________
Suite Number____________
Street____________
City____________
County____________
State____________ Zip Code____________

5. Age and Condition____________

____

6. Landlord or Agent____________
Suite Number____________
Building____________
Street____________
City____________
State____________ Zip Code____________
Telephone Area Code______ Number____________

7. Current Occupant of Space____________

8. Current Use of Space____________

9. Term of Lease____________
From____________ To____________

10. Renewal Options____________

____

____

Figure 2-8. Preliminary Investigation Report.

Facilities Department will be in a position to select the site best suited to the need at a cost consistent with the original intent of the space.

The second essential form is the Preliminary Investigation Summary. This, as the name implies, provides a summary of the information contained in the Preliminary Investigation Report. Please note that the Summary does not specifically identify any of the sites investigated and recorded in the Investigation Report. There is a reason for

PRELIMIARY INVESTIGATION REPORT

11. Options on Additional Space____________________

12. Right of First Refusal on Additional Space____________________

13. Area in Square Feet

    Rentable__________

    Usable__________

14. Rental Costs

    Square Foot Per Year $__________

    Square Foot Per Month $__________

    Total Per Year $__________

    Total Per Term of Lease $__________

15. Escallations

    Building Operating Expenses__________

    Real Estate Taxes__________

    Other__________

16. Lease Clauses

    Right to Sublet____

    Early Terminations: When__________

    Penalty__________

    Obligations for Restorations $__________

    Abatement of Rent for Extreme Casualties for__________Days

    Other____________________

Figure 2-9. Preliminary Investigation Report.

this absence of specific, detailed information relative to any one site. The intent of the Summary is to obtain corporate approval for the Facilities Department to proceed with lease negotiations with as many owners as good judgment dictates. Often, in the final analysis, depending upon the outcome of the lease negotiations, any one of several sites may be selected.

The Preliminary Investigation Summary serves its purpose in that it notifies top management of the following specific information:

- Unit requesting space
- Location of space
- Reason for request
- Monthly, annual, and total lease liability
- Duration of lease

PRELININARY INVESTIGATION REPORT

17. Building Services & Utilities Furnished by Landlord

Heating: From________ To________ Daily

From________ To________ Annually

Air Conditioning: From________ To________ Daily

From________ To________ Annually

Janitorial Service____

Window Cleaning_______

Other________

18. Building Services Furnished by the Company

Description	Annual Cost
________	$________
________	________
________	________
________	________

19. Leasehold Improvement Allowance by Landlord

Partitions________

Doors________

Door Hardware________

Ceiling________

Lighting________

Floor Covering________

Wall Finish________

Electrical Outlets (Wall)________

Electrical Outlets (Floor)________

Telephone Outlets (Wall)________

Telephone Outlets (Floor)________

Window: Blinds________

Drapes________

Other________

____

Figure 2-10. Preliminary Investigation Report.

This document is the vehicle that puts into motion the lease negotiations. In the event that there are any changes in plans for this action, the same document can be used to either abandon or redirect the leasing effort.

You are now ready to consumate the lease. To assist you with this portion of your work, we have included a *typical commercial lease.* While we call this a typical commercial lease, such an instrument does not exist. What we are trying to do is to expose you to the many and various terms included in a typical lease so that you will be in the

PRELIMINARY INVESTIGATION REPORT

20. Estimated Leasehold Improvement Costs Above Building Allowance

____

____

21 Parking: Describe______________________________

____

____

Number of Parking Spaces Included in Lease______________

Estimated Annual Cost for Additional Parking Spaces:

$______________ Per Year______________ Per Space

22. Current Use of the Building______________________

____

____

23. Comments______________________________________

____

____

Prepared By ______________________________

____

______________________________ Date______________

Approved By ______________________________

____

______________________________ Date______________

Figure 2-11. Preliminary Investigation Report.

best possible posture in dealing with leases. What follows is a guide; information to familiarize you with typical terms. In the event that you have a legal staff, have them review the lease. This is legal stuff, treat it as such.

TYPICAL COMMERCIAL LEASE

There are about as many different types of leases as there are landlords, and the complexity of the leases increases in direct proportion to the amount of space, time, and money involved. As you would imagine, a month-to-month lease is *as-is* space, without leasehold improvements is probably the simplest. There are very few restrictions and stipulations, and most likely no options. You, the tenant, can use the space, until terminated by either party, usually with 30 days written notice.

At the other extreme, a long-term lease agreement involving a significant amount of space and money is completely different. Some landlords, with their own legal staff, may even have a lease designed for that specific property. Others may use standard

PRELIMINARY INVESTIGATION SUMMARY

1. Organization Requesting Space______________________________

2. Unit Manager______________________________

3. Location of Space, City________________ State________________

4. Use of Space______________________________

____

5. Reason for Request, New____ Renewal____ Relocation____ Expansion___

Consolidation____ Other____

6. Term of Lease______________________________

7. Space Description, Area of Rentable Space________________ Sq. Ft.

Reserves Parking Space________________

Monthly Rent $________________

Annual Rent $________________

Total Lease Liability$________________

Annual Unit Cost $________________/Sq. Ft.

8. Major and/or Unusual Lease Provisions________________

____

9. Estimated Leasehold Improvement Costs Above Building Allowance, $_____

10. Population, Present________________

At Term of Lease______________

11. Comments______________________________

____

Prepared by________________________

Date________________________

Leasing Activity Approval________________________

Date________________________

Figure 2-12. Preliminary Investigation Summary.

lease agreement forms. In any case, lease agreements usually, enumerate the conditions, provisions, and options that the landlord (lessor) imposes upon and offers to the tenant (lessee). Some contain rules and regulations established by the landlord that form a part of the lease and are binding upon both parties.

Leases also contain additional documentation riders, attachments, or exhibits such as rules and regulations, description of landlord's work, building standards, work letter, unit prices for extras and credits, and drawings.

The following is a list of articles found in typical commercial leases:

1. Parties
2. Premises
3. Use
4. Term and Possession
5. Rent
6. Security Deposit
7. Taxes
8. Expenses
9. Place of Payment
10. Inability to Give Possession
11. Landlord's Services
12. Rights Reserved to Landlord
13. Holding Over as Renewal
14. Quiet Enjoyment
15. Waiver of Subrogation, Indemnification, and Public Liability Insurance
16. Fire Insurance
17. Negative Covenants of Tenant
18. Signs
19. Additions and Improvements
20. Vacation
21. Condemnation
22. Damage or Destruction
23. Assignment or Subletting
24. Performance of Tenant's Covenants
25. Tenant Default
26. Tenant's Option to Renew
27. Tenant's Option to Contiguous Space
28. Notice of Termination
29. Lease Contains All Agreements
30. Attachments and Exhibits
    A. Rules and Regulations
    B. Landlord's Building Standard Work
    C. Building Standard Unit Prices, Extras, and Credits
    D. Cleaning Specifications
    E. Drawings

From this it would appear that lease agreements are complex documents, and they are, in many cases. Frequently, the major problem is that lease agreements are written by lawyers for lawyers. Fortunately, some are written by lawyers for the layperson. Sometimes, even common-term agreements are difficult to understand because the language varies from lease to lease. For example, see *Parties* as described below in Lease 1, Lease 2, and Lease 3.

As an example, we have selected the opening statement of three different leases from different states under the common heading *Parties*

Parties

Lease 1. THIS INSTRUMENT IS AN INDENTURE OF LEASE in which the Landlord and the Tenant are the parties hereinafter named, and which relates to space in the building (the Building known as BIG EXECUTIVE CENTER . . .)

Landlord: ____________________

Tenant : ____________________

Lease 2. THIS AGREEMENT of Lease, made this ______ day of __________, 19 , between BIG CORPORATION, a corporation, authorized and doing business under and by virtue of the laws of the Commonwealth of Big State (for the purpose of this Lease to be known as the "Lessor"), and Small Insurance Company, Inc., a Corporation, authorized and doing business under and by virtue of the laws of the Commonwealth of Big State (for the purpose of this Lease to be known as the "Lessee") . . .

Lease 3. THIS LEASE, made this ______ day of __________, 19 , by and between Big Building Co., hereinafter called "Landlord," and Small Insurance Co., hereinafter called "Tenant."

As you can see there are several different ways to identify the parties to a lease. The same applies to the entire lease agreement. We will attempt, with some degree of success, we hope, to take you through a typical commercial lease.

Premises

WITNESSETH, that the Landlord hereby rents to the Tenant, and the latter does hereby rent from the former, all that certain space of ________ square feet, to be used for office purposes, in the building known as ____________ located on ____________ in ____________. Said premises shown on drawing(s) marked Exhibit E, attached hereto and made a part hereof, and to be substantially completed in accordance with the basic specifications shown on Exhibit B attached hereto, and made a part hereof, for the term of ( ) years, beginning on the commencement date as hereinafter defined.

Also, a minimum of __________ parking spaces located ____________ the building.

This article should describe the use of the premises without restricting the intended use and beware of such clauses as *and for no other purpose.* The following is an example:

Use

Lessee shall use and occupy Demised Premises for office purposes, including executive dining room, operation of equipment and facilities of computer service, and reproduction of written, printed, and photographic materials for internal use by Lessee and any permitted Sublessees, and for no other purpose.

Term and Possession

The term of this lease shall begin on the ______ day of __________, 19 , and end on the ______ day of __________, 19 , (unless further extended or sooner terminated as hereinafter provided).

The Commencement Date, whenever used herein, shall be __________, or if no date certain is specified, the first day of the first calender month on or following the first to occur of the following: (a) the date when the premises are ready for occupancy; or (b) the date when the Tenant shall take possession and occupy the premises. The premises shall be deemed ready for occupancy when the work to be done by Landlord is substantially completed or would have been except for delay due to Tenant, and a Certificate so stating is issued and signed by Landlord and mailed in an envelope with postage affixed and deposited in the United States Mail and addressed to the Tenant at ________________. In the event of any dispute as to when said Landlord's work has been or would have been substantially completed, as aforesaid, the decision of the Landlord's architect on said issues shall be final and binding on the parties.

Rent

For the use of said premises, the Tenant agrees to pay the Landlord, in equal monthly installments of ____________ ($________), the first installment is due and payable on the commencement date, with subsequent installments due and payable on the first day of each calender month thereafter until a total rental of ($________), shall be paid. In the event Tenant shall occupy the premises prior to the commencement date, Tenant shall pay, in addition to all other sums specified, for each day, including the first day of occupancy, to the commencement hereof, which payments shall be made on the commencement date. The Tenant agrees to pay concurrently with the signing of this lease the sum of __________ ($________) to be credited to the first month's rent.

The Tenant agrees to pay the rent herein provided without deductions whatsoever, and without any obligation on the Landlord to make demand for it. Any installment of rent accruing hereunder and any other sum payable hereunder, if not paid when due, shall bear interest at the rate of ______ percent (%) per annum until paid. Time is of the essence of this lease.

Security deposit

As security for the faithful performance by Tenant of all the terms and conditions upon Tenant's part to be performed and for the payment of any damages to which Landlord may be entitled in the event of default by Tenant under the Lease, Tenant has this day deposited with Landlord the sum of ($________) which shall be returned to Tenant.

Taxes

The Landlord shall pay all real estate taxes. . . .

or

Taxes

The Tenant shall pay, as additional rent, its proportional share, as hereinafter determined, of any increase in real estate taxes for any future lease year. . . .

Expenses

The Tenant shall pay, as additional rent, its proportional share, as hereinafter determined, of any increase in expenses paid or incurred by the Landlord for maintaining, operating, and repairing the real property, and the cost of electricity, water, fuel, all insurance, and any and all other expenses which would be considered as an expense of maintaining, operating, or repairing the property.

Place of Payment

The rent shall be paid without notice or demand, and without abatement at the principal office of the Landlord.

Inability to Give Possession

The Landlord shall not be liable to the Tenant for any damage which may be caused to the Tenant by the failure of the Landlord to give possession of the demised premises at the time agreed upon, if said failure is not due to any fault on the part of the Landlord.

Landlord's Services

The Landlord shall: provide elevator facilities on business days from ______ A.M. to ______ P.M. and on Saturdays from ______ A.M. to ______ P.M. and have an elevator subject to call at all other times; furnish heat to the demised premises, when necessary, on business days from ______ A.M. to ______ P.M. and on Saturdays from ______ A.M. to ______ P.M.; furnish air conditioning to the demised premises, when necessary, from May 1 to September 30, on business days from ______ A.M. to ______ P.M. and on Saturdays from ______ A.M. to ______ P.M.; clean or cause the demised premises to be kept clean; provided electrical service; provide hot and cold water for ordinary cleaning, toilet, lavatory, and drinking purposes; provide snow and ice removal from pavement(s), driveway(s) and parking area(s). (Unless cleaning is further defined under Rules and Reglations, it must be addressed here. The lease usually includes a clause stating that the Landlord shall have no responsibility or liability for failure to provide services by any cause beyond Landlord's control.)

Rights Reserved to Landlord

The Landlord reserves the right to display a "For Rent or For Sale" card or sign upon premises, and to show same to prospective tenants or buyers. The Landlord may enter the premises at reasonable times during the term, for the purpose of inspecting. The Landlord may from time to time establish reasonable rules and regulations for the safety, care, and cleanliness of the premises, and for the preservation of good order therein.

Holding Over as Renewal

A holdover by the Tenant beyond the term of this lease shall be deemed a renewal of the lease on a year-to-year term basis, with said renewal being under and subject to all provisions contained in this Agreement of Lease, and shall be at the option of the Landlord.

Quiet Enjoyment

The Tenant, upon paying the said rent and performing the covenants of this lease, shall and may peaceably and quietly have, hold, and enjoy the demised premises for the term aforesaid and herein duly authorized additional term.

Waiver of Subrogation, Indemnification, and Public Liability Insurance

The Tenant shall not do or permit any act or thing upon the demised premises which may subject the Landlord to any liability or responsibility for injury, damages to persons or property . . . and shall fully protect Landlord against such liability. Tenant shall indemnify and hold harmless Landlord from and against any and all liability, fines, suits, claims, demands. . . .

The Tenant shall keep in force at its own expense public liability insurance in companies acceptable to Landlord sufficient to cover such indemnification.

The Landlord and Tenant hereby release the other from any and all liability or responsibility to the other or anyone claiming through or under them by way of subrogation or otherwise for any loss . . . (indicates further inclusions of definitions)

Fire Insurance

The Landlord shall insure the building on the demised premises with fire and extended coverage insurance in the amount of at least 80 percent of the value thereof; provided, however, that the Landlord shall not be required to insure improvements or additions placed in the demised premises by the Tenant.

Negative Covenants of Tenant

No waste, damage, or injury to the premises shall be committed, and at the end of the term the demised premises shall be restored, at the option of the Landlord, to the same condition in which they were at the commencement of the term, and the cost of said restoration shall be paid by the Tenant. This paragraph is subject to the exceptions of ordinary wear and tear and unavoidable damage by fire, elements, casualty, or other cause or happening not due to Tenant's negligence.

The Tenant shall not carry on any unlawful or immoral business in or about the demised premises, and shall not carry on any business which will endanger the building from fire or cause a forfeiture of any fire insurance that the Landlord has or may hereinafter have on said building. The Tenant shall not operate any machinery or equipment that may be harmful to the building or disturbing to other occupants of the buildings. The Tenant shall not place weights in any portion of the demised premises beyond the safe carrying capacity of the structure.

Signs

No showcase, sign, or hanging or protruding sign or permanent obstruction of any kind shall be kept or maintained by the Tenant on the reservation or sidewalk in front of the demised premises, said space to be used only for purposes of ingress and egress. If the Tenant desires to affix any exterior sign(s) on the building or premises, etc., the same may be affixed, operated, and maintained, at the expense of the Tenant, upon securing written permission from the Landlord.

Alterations and Improvements

The Tenant shall not have the privilege of improving the leased premises for Tenant's own purposes, unless the Landlord approves any proposed improvements in writing, and improvements are made at Tenant's expense.

Vacation

The Tenant shall neither vacate or desert the premises during the term of the lease nor permit same to the empty or unoccupied.

Condemnation

In the event the whole or any part of the leased premises shall be taken under the power of eminent domain, or sold under threat thereof, or taken in any manner for public use, the Landlord, at its option, may terminate this lease, which lease shall then terminate on the effective date of the condemnation or sale. The compensation awarded or paid for such taking, both as to Landlord's reversionary interest and Tenant's interest under this lease, shall belong to and be the sole property of the Landlord. Tenant shall have no claim against the Landlord or be entitled to any award or damages other than an abatement from the rent beyond the period of termination date and compensation paid for moving expenses and/or costs of removal of stock and/or trade fixtures.

Damage or Destruction

If the premises are partially damaged by fire or other casualty, not the fault of the Tenant, Landlord shall make repairs as speedily as conveniently possible. If the damage is so extreme as to render the premises wholly unfit for occupancy, the rent shall cease until the premises are put in repair. In the event of total destruction of the building of which the demised premises form a part, or if in the judgment of the Landlord the damage to the demised premises cannot be repaired within one hundred twenty (120) days, and if the Landlord shall decide not to restore or repair same, or shall decide to demolish the building, the Landlord may, within ninety (90) days after such fire or other casualty, by notification to the Tenant, terminate this lease. Tenant shall continue to pay the rent in the event the premises are only partially damaged and fit for occupancy. In no event shall Landlord be liable for any loss or damage sustained by Tenant by reason of fire or other accidental casualty.

Assignment or Subletting

Tenant shall not assign, mortgage, or encumber this lease, nor sublet the premises or any part thereof. In the event of the insolvency or bankruptcy of Tenant, this lease shall, at the option of the Landlord, terminate forthwith, and this lease shall not by operation of law or otherwise, be considered a part of Tenant's estate.

Performance of Tenants Covenants

The Tenant shall perform all agreements herein expressed on its part to be performed, and it will promptly upon receipt of written notice of nonperformance thereof, comply with the requirements of such notice; and further, if Tenant shall not comply with such notice to the satisfaction of Landlord within a reasonable time after notice except in case of emergency, Landlord may, at its option, do or cause its agents, employees, and contractors to enter upon the demised premises and in such event shall have no liability to Tenant for any loss or damage resulting in any way from such action; the Tenant shall pay promptly upon demand any expense incurred by Landlord in taking such action as additional rent.

Tenant Default

In the event of Tenant default, as described hereinafter, the Landlord shall have the right to enter the premises and distrain for any amount of money that may be due under this lease, either by force or otherwise, without being liable to any prosecution therefore, and to apply any proceeds to the payment of the rent due, holding the Tenant liable for any deficiency. The occurrence of any of the following shall constitute an event of default:

a. Failure of Tenant to take possession of the demised premises within ______ (   ) days after commencement date of this lease.

b. Nonpayment of rent by Tenant within ten (10) days after notice by Landlord.

c. Discontinuance by Tenant of the conduct of its business in the demised premises.

d. Vacation or desertion of the demised premises or permitting the same to be empty or unoccupied.

e. Removal or attempt to remove goods or property from or out of the demised premises otherwise than in the ordinary and usual course of business without having first paid or satisfied Landlord for all rent which may become due during the entire term of this lease.

f. Tenant's failure to vacate and surrender the premises as required in this lease upon expiration of the term or termination of this lease.

g. Tenant's failure to perform any other agreement or condition of this lease within ten (10) days after written notice and demand, unless the failure is of such a character as to require more than ten (10) days to cure, in which event Tenant's failure to proceed diligently to cure such failure shall constitute an event of default.

h. Tenant's bankruptcy or insolvency.

Tenant's Option to Renew

Tenant shall have the option to renew this lease for an additional term of _______ (   ) years upon written notice to the Landlord received at least four (4) months prior to the end of the term herein set forth.

Tenant's Option to Continguous Space

Landlord shall give first option to Tenant to lease contiguous space when such space becomes available.

Notice of Termination

Anything heretofore notwithstanding, a written notice that the Tenant intends to terminate this lease shall be delivered to the Landlord at least four (4) months prior to the end of the term of this lease or the Tenant's inaction shall result in the renewal of this lease for a further term of one (1) year and so on from year to year, and the said renewal(s) shall be under and subject to all provisions as contained in this Agreement of lease; provided, however, that such renewal(s) shall be at the option of the Landlord, and shall be at the rental rate then in effect for the building.

Lease Contains all Agreements

It is expressly understood by the parties that the whole agreement is embodied in this Agreement (executed in duplicate) and that no parts or items are omitted, unless the same be hereinafter modified by written agreement(s).

This Agreement shall be binding upon the heirs, executors, administrators, successors and/or assigns of the parties hereto.

Reference to masculine, feminine, or neuter gender shall include proper gender as the case may be.

Reference to singular or plural shall be deemed singular or plural whenever the sense of the instrument so requires.

If the subject matter of a provision of the Rules and Regulations is discussed in a provision in the body of the lease, both provisions shall be effective as far as the same remain consistent; however, if the provisions are contradictory, then the provisions in the body of the lease shall control.

IN WITNESS WHEREOF, the parties hereto, by the properly authorized persons and with the respective seals attached, have fully executed this lease the day and year first written above.

Witness: BIG EXECUTIVE CENTER

________________ By ________________

Landlord

BIG CORPORATION

________________ By ________________

Tenant

Exhibits A through D are typical attachments to leases.

EXHIBIT A
RULES AND REGULATIONS

1. The sidewalk, halls, passages, elevators, and stairways shall not be obstructed by any of the tenants, or used by them for any other purpose than for ingress and egress to and from their respective leased premises.

2. Tenants, their clerks, or servants, shall not make or commit any improper noises or disturbances of any kind in the building, or mark or defile the water-closets, toilet rooms, walls, windows, elevators, or doors of the building, or interfere in any way with other tenants or those having business with them.

3. No carpet, rug, or other article shall be hung or shaken out of any window or placed in corridors as a door mat, and nothing shall be thrown or allowed to drop by the tenants, their clerks, or employees, out of the windows or doors, or down the passages or shafts of the building, and no tenants shall sweep or throw, or per-

mit to be thrown from the leased premises, any dirt or other substance into any of the corridors or halls, elevators, shafts, or stairways of said building.

4. The toilet rooms, water-closets, and other water apparatus shall not be used for any purpose other than those for which they were constructed, and no sweeping, rubbish, rags, ashes, chemicals, or the refuse from electric batteries, or other unsuitable substances, shall be thrown therein. Any damage resulting from such misuse or abuse shall be borne by Tenant by whom or by whose employees or visitors it shall be caused.

5. No linoleum, or oil cloth, or rubber, or other airtight coverings shall be laid on the floors, nor shall articles be fastened to or holes drilled or nails or screws driven into the walls, windows, partitions, nor shall the walls or partitions be painted, papered, or otherwise covered, or in any way marked or broken, without the prior written consent of the Landlord.

6. Nothing shall be placed on the outside of the building, or on the windows, windowsills, or projections.

7. No sign, advertisement, or notice shall be inscribed, painted, or affixed on any part of the outside or inside of said building unless of such color, size, and style, and in such places upon or in said building as shall be first designated by Landlord. Signs on doors and windows, where permitted will be affixed for the Tenants by a contractor chosen by Landlord, the cost of the sign and of the affixing to be paid by Tenant. Directory in the lobby, with the names of Tenants, will be provided by Landlord in accordance with Rule 22 hereof.

8. After permission to install telephones, call boxes, telegraph wires, or other electric wires has been granted, Landlord will direct where and how the same are to be placed. No wires shall be run in any part of the building excepting by or under the direction of Landlord. The attaching of wires to the outside of the building is absolutely prohibited. No boring or cutting of floors or partitions for wires is permitted except with prior consent of Landlord.

9. Tenants may use their own safes, but Landlord shall have the right to prescribe the weight and proper position of safes, and no safe shall be hoisted or placed in any part of the building excepting under the direction of Landlord's agents. All damage to the building caused by installing, maintaining, or removing safe, furniture, equipment, or other property shall be repaired at expense of Tenant.

10. Each Tenant will be furnished one or more keys to the premises, and must, upon termination of his lease, surrender all keys delivered to said Tenant.

11. No tenant shall do or permit anything to be done in said premises, or bring or keep anything therein which will in any way increase the rate of fire insurance on said building or on property kept therein, or obstruct or interfere with the rights of other tenants, or in any way injure or annoy them, or conflict with the laws relating to fires, or with the regulation of the Fire Department or with any insurance policy upon said building or any part thereof, or conflict with any of the rules and ordinances of the Department of Health. Tenant shall not conduct or permit any auction on the premises.

12. The Landlord will provide janitorial service to perform the following general cleaning:

Sweep all tile or similar type flooring, sweep or vacuum clean carpeting or rugs, and dust portions of furniture, fixtures, and windowsills, empty and remove waste materials from normal office waste receptacles, empty and clean ashtrays, and dust doors and ventilating louvers.

Landlord shall have no obligation for cleaning the leased premises other than as specifically set forth herein.

13. No tenant shall employ any person or persons for the job of cleaning premises unless said cleaning and the person or persons have first been approved in writing by Landlord. Tenants will see that windows are closed and the doors securely locked before leaving the building.

14. No animals or birds shall be brought into or kept in or upon the premises.

15. No machinery of any kind shall be allowed to be operated on the premises without the written consent of Landlord.

16. No interference with the heating apparatus will be permitted. All regulating and adjusting will be done by employees of Landlord.

17. The use of office suites as sleeping apartments, or for any immoral or illegal purpose is absolutely prohibited.

18. Landlord reserves the right to make any or all alterations in premises as may be required by Tenant, the expense of such alterations to be paid by Tenant.

19. No furniture or other material shall be moved into or out of the building without first notifying the superintendent, and the moving thereof shall be under his direction and control.

20. Any and all damage to floors, walls, or ceiling or to personal property due to Tenant or Tenant's employees' failure to shut off running water in any piece of equipment shall be paid by Tenant.

21. Tenant shall not place a load upon any floor of the demised premises exceeding the floor load per square foot area which such floor was designed to carry and which is allowed by law. Business machines and mechanical equipment shall be placed and maintained by Tenant at Tenant's expense in settings sufficient, in Landlord's reasonable judgment, to absorb and prevent vibration, noise, and annoyance.

22. The Directory Board in the entrance lobby of the Building is provided for the exclusive display of the name and location in the Building of each tenant. Each tenant shall pay the cost of placing tenant's name on the Directory Board. Landlord reserves the right to exclude from the Directory Board any name other than Tenant's name. If Landlord, at Landlord's option, agrees to permit other names to be added to the Directory Board, Landlord shall have the right to levy a reasonable rental charge for the same.

23. Landlord reserves the right to exclude or expel from the Building any persons who, in the judgment of Landlord, are intoxicated or under the influence of liquor or drugs, or who shall do anything in violation of the Rules and Regulations of the Building.

24. Room used in common by Tenant shall be subject to such regulations as are posted therein.

25. Landlord reserves the right to close and keep locked all entrance and exit doors of the Building during hours Landlord may deem advisable for the adequate protection of the property. Use of the Building and the leased premises before 8:00 A.M. or after 6:00 P.M., or at any time during Saturday, Sunday, or legal holidays shall be permissive, and subject to the rules and regulations Landlord may prescribe.

The Tenant and its employees, agents, and associates, or other persons entering or leaving the Building at any time, when so locked, may be required to sign the Building register, and the watchman or Landlord's agent in charge shall have the right to refuse admittance to any person into the Building without a pass, or other satisfactory identification, showing right of access at such time. Landlord assumes no responsibility and shall not be liable for any damages resulting from the admission of any authorized or unauthorized person to the Building.

26. Landlord shall have the right to prohibit any advertising by any tenant which, in Landlord's opinion, tends to impair the reputation of the Building or its desirability as a building for offices, and upon written notice from Landlord, such tenant shall refrain from or discontinue such advertising.

27. No tenant will install blinds, shades, awnings, or other form of inside or outside window covering, or window ventilators or similar devices, without the prior written consent of Landlord.

28. Tenant and its agents and employees shall be subject to such rules and regulations regarding the parking facilities as may be promulgated by the Landlord. Landlord reserves the right to change the location, layout, and arrangement of the parking areas and driveways and to restrict parking by Tenant, its agents, and employees to areas designated by Landlord.

29. Positively no carpet may be cemented to the concrete floor.

EXHIBIT B
LANDLORD'S BUILDING STANDARD WORK

The Landlord agrees to perform, at its expense, the following work (Landlord's Work) in Building Standard manner and material in the demised Premises. The Tenant agrees to submit its plans for its demised Premises on or before ________ ________.

Ceiling: 12″ × 12″ × $\frac{3}{4}$″ mineral fissured acoustic tile on a concealed spline system.

Flooring: 9″ × 9″ × $\frac{1}{8}$″ marbleized vinyl tile with straight or coved base.

Lighting: 2′ × 4′4 lamp, 40-watt fixture on a basis of one per 66 square feet, net usable area.

Partitions: 1 lineal foot per 12 square feet net usable area of stud and Sheetrock partition consisting of 2-$\frac{1}{2}$″ steel stud, 3 layers of $\frac{1}{2}$″ Sheetrock (one on one side, two on the other side) or 2-$\frac{1}{2}$″ stud with one layer of $\frac{5}{8}$″ Sheetrock each side and fiberglass insulation packed between the studs.

Doors, Bucks, and Hardware: One full height hollow metal door per 30 lineal feet of partition, together with 16-gauge hollow metal buck and latchset butts, doorstop per opening. Tenant to receive one lockset and one door closer per tenant or per floor.

Paint: All partitions, doors, and bucks to receive two or three coats (as required to cover) on all Building Standard partitions, doors and bucks, one color per room, light tones.

Electric: Based on 5 watts (3 lighting, 2 power); one 15-amp 120-volt floor outlet (on duct pattern), or wall outlet per 100 square feet, net usable area; one floor telephone outlet (on duct pattern) or in wall per 150 square feet net usable area; one switch per room or per bay if open area.

Window Shading: Draperies (standard of the building) on all windows.

Heating and Air Condtioning: Perimeter zone including induction unit covers in Base Building Contract. Provide sufficient distribution ducts and outlets to service tenant areas constructed in accordance with above standards.

EXHIBIT C
BUILDING STANDARD UNIT PRICES, EXTRAS, AND CREDITS

The Landlord agrees to perform, at the Tenant's expense, in Building Standard manner and material in the demised Premises, any additional work (over Landlord's Building Standard Work, EXHIBIT B) as required by the Tenant. Landlord agrees to give credit for any work not required by the Tenant according to EXHIBIT B.

ITEM	*EXTRA*	*CREDIT*
1. Partition	__________	__________
2. Door	__________	__________
3. Latchset	__________	__________
4. Lockset	__________	__________
5. Light fixtures	__________	__________
6. Duplex base receptacles	__________	__________
7. Floor receptacles	__________	__________
8. Telephone wall outlet	__________	__________
9. Telephone floor outlet	__________	__________
10. S/P switch	__________	__________
11. 3/W switch	__________	__________
12. Resilient floor tile	__________	__________
13. Carpet	__________	__________
14. Vinyl wall covering	__________	__________
15. Building standard painting	__________	__________

EXHIBIT D
CLEANING SPECIFICATIONS

Applicable only to Office Area and in
Part to Certain Public Areas

GENERAL

All stone, ceramic tile, marble, terrazzo, and other unwaxed or untreated flooring to be swept nightly, using approved dust-down preparation; wash such flooring once a month.

All linoleum, rubber, asphalt tile, and other similar types of flooring (that may be waxed or treated) to be swept nightly, using approved dust-down preparation. Waxing, bimonthly, shall be done at Landlord's expense.

All carpeting and rugs to be carpet swept nightly and vacuum cleaned once each week.

Hand dust and wipe clean all furniture, fixtures, windowsills, and convector closure tops nightly; wash sills and tops when necessary.

Empty, clean and damp dust as necessary all waste receptacles nightly and remove from the demised premises waste paper and waste materials.

Empty and clean all ashtrays and screen all sand urns nightly.

Wash clean all water fountains and coolers nightly.

Dust all door and other ventilating louvers within reach, as necessary.

Dust all telephones as necessary.

Sweep all private stairway structures nightly.

Keep locker and slop sink rooms in a neat and orderly condition at all times.

Wipe clean all brass hardware and brass fixtures and other brightwork nightly.

Check all stairwells throughout entire building daily and keep in clean condition.

Properly maintain metal doors of all elevator cars.

Wipe all interior metal window frames, mullions, terrace doors, and other unpainted interior metal surfaces of the perimeter walls of the building each time the interior of the windows is washed. Such surfaces to be cleaned once each year with Calmal or other appropriate cleaner.

Vacuum clean peripheral induction air-conditioning units semiannually.

LAVATORIES

Sweep and wash all lavatory floors nightly, using disinfectants. Wash and polish all mirrors, powder shelves, brightwork, and enameled surfaces in all lavatories, nightly.

Scour, wash, and disinfect all basins, bowls, and urinals nightly.

Wash both sides of all toilet seats nightly.

Hand dust and clean, washing where necessary, all partitions, tile walls, dispensers, and receptacles in all lavatories and restrooms nightly.

Fill toilet tissue holders nightly (tissue to be furnished by Landlord).

Empty sanitary disposal receptacles nightly.

Wash interior of waste cans and receptacles at least once a week.

Wash and polish all wall tile and stall surfaces as often as necessary but in no event less than once every two weeks.

Fill soap dispensers and paper towel dispensers.

Empty paper towel receptacles and transport wastepaper from the demised premises, nightly.

BUILDING LOBBY, ETC.
Floor of entrance lobby to be swept and washed nightly.

Floors in elevator cabs to be washed and waxed and polished nightly—walls and metal work and saddles to be dusted and rubbed down nightly.

Wipe down mail chute and mail depository nightly.

Lobby walls to be dusted as often as necessary and washed down when necessary.

Clean lights, globes, and fixtures as often as necessary throughout the lobby.

Screen and clean sand urns nightly.

HIGH DUSTING
Do all high dusting quarterly, which includes the following:

Dust all pictures, frames, charts, graphs, and similar wall hangings not reached in nightly cleaning.

Dust clean all vertical surfaces, such as walls, partitions, doors and bucks, and other surfaces not reached in nightly cleaning.

Dust clean all pipes, ventilating and air-conditioning louvers, ducts, high moldings, and other high areas not reached in nightly cleaning.

Dust all lighting fixtures, including plastic and glass enclosures.

Dust all venetian blinds.

WINDOW CLEANING
All windows from the first floor to the roof to be cleaned inside and outside approximately three times a year.

Public entrance doors and lobby glass to be cleaned daily and kept in clean condition at all times during the day.

All interior glass and a normal amount of partition glass, glass doors, and fan lights are to be cleaned approximately once every five weeks.

Mail chute glass to be kept in clean condition.

DAY PORTERS
Necessary regular porters will be assigned to the building for the following services:

To service all public and operating space throughout the building.

To check condition of elevator cars during the day.

To dust and rub down all elevators doors, frames, telephone booths, directory, etc., daily.

To sweep sidewalks and ramps daily.

To hose sidewalks, weather permitting.

To remove snow when necessary.

To clean roof setbacks as often as necessary.

To lay and remove lobby rubber runners as necessary.

To insert toilet tissue in lavatories (tissue to be furnished by Landlord).

To keep fan rooms, motor rooms, and air-conditioning rooms in clean condition.

To check condition of staircases during the day.

To maintain exterior granite; exterior metalwork along with metal in building entrance doors and storefront trim. To clean standpipes and sprinkler siamese connections and hose bibs.

All cleaning services except those performed by day porters and window cleaners are to be performed nightly, five nights per week. No Saturday, Sunday, or legal holiday service—legal holidays to be those stated in the building service union agreements.

The contractor or Landlord is to furnish all necessary approved cleaning material, implements, and machinery for the satisfactory completion of the work. This includes escalator platforms, vacuum machines, and scrubbing machines.

The contractor or Landlord will furnish uniforms for all of its employees.

SUMMARY

A systematic approach to leasing property will yield the most satisfactory results. The following includes all considerations such as, the organization chart, space and furniture standards, etc., which you may already know about and have.

1. Develop an organization chart.
2. Develop space and furniture standards.
3. Develop a leasing procedure.
4. Determine the type of lease required.
5. Determine the amount of space required.
6. Conduct a lease investigation and comparison of several sites.
7. Compare rental costs.
8. Develop a lease agreement (seek professional assistance when necessary).
9. Complete lease agreement.

3. Comparing Rental Costs

Before signing any lease, the several sites investigated must be evaluated. Comparing rental costs and determining which property provides the greatest amount of space for the least amount of money may well be the most difficult phase of the entire site search and leasing operation. Also, this is the stage at which you may be committing your company to a significant, long-term investment. A false move here could have a drastic effect on the company's profit structure.

It is important to become familiar with the various terms commonly used in building construction and real estate leasing in order to be in a position to compare rental costs intelligently. All of the economical comparison problems would disappear if, and that's a big if, all landlords used the same terms to define the space described in a particular lease, and if that term is consistently applied to such space.

COMMON TERMS

Most leases use the term *rentable area* to define the space under consideration, some leases describe the area in square feet, and others provide a drawing showing the area involved. Some leases include the cost for the space in dollars per square foot on an annual basis, while others state the cost of the space on a monthly or annual dollar basis.

It is inadequate to compare the cost in dollars per square foot of one building against the others because there are entirely too many variables involved in any leasing situation. First of all, there is no simple uniform method for determining the true cost of space. This is further complicated by the fact that some leases are written to include all costs for the space while others, in addition to a base cost for space, have additional charges for maintenance and operation, utilities, real estate taxes, and escalations. It could be a very simple matter to add up all costs involved with the lease and divide it by the space described to obtain a true cost. Unfortunately, it is not quite that easy. It is even difficult to determine the actual space in every situation. The best place to start the analysis for comparing rental costs is with a typical building.

Gross Area

Figure 3-1 is a typical floor plan of a high rise office building. This plan includes several essential dimensions, for example, the core area is 40 × 80 feet and is rather straightforward, but there are two dimensions along the exterior of the building. The gross area of this floor, using the outside dimensions of the building, is 25,900 square

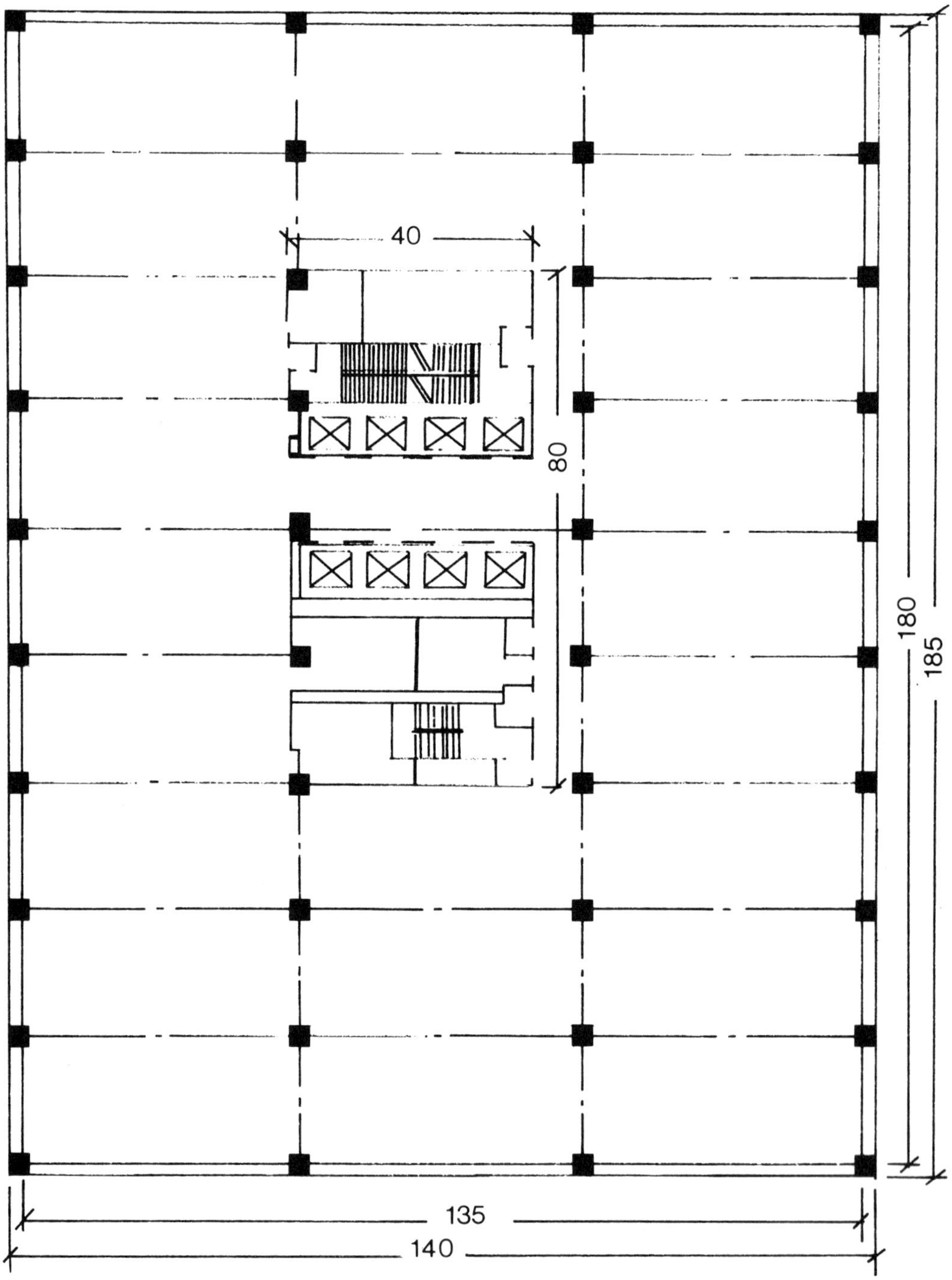

Figure 3-1. Typical Floor Plan of a High Rise Office Building.

feet, but the total gross area of the building is not the number obtained by adding together all tenant floors. It must include space that is not considered tenant space.

This building has a full basement of mechanical space, and one entire intermediate floor is also devoted to mechanical equipment. According to the architect who designed the building, the gross area of the building consists of all spaces calculated using outside dimensions of perimeter walls. Using this method of determining gross area provides a means for comparing costs on a dollars per square foot. This is a valuable tool for comparing construction costs. Merely stating that a building costs $30 million to construct is important only in so far as it can be used to establish total cost, but describing the building as costing $80 per square foot readily defines it. In fact, A/E firms, constractors, building owners, lending institutions, and real estate agents all use these figures to evaluate building construction costs.

Usable Area

Chapter 2 showed how to determine the amount of usable area required, and this is the information you use in seeking a site. The usable area in Figure 3–1 is significantly less than the gross area; in fact, it is only 21,100 square feet. In this case, the perimeter walls and the core area, 40 × 80 feet, are excluded.

Usable area is sometimes defined as "any area where furniture can be placed." In the purist sense, this is not a true statement. All interior partitions were included in the development of usable space in chapter 2. From this, it is obvious that a 10 × 15-foot office is nominally 10 × 15 feet, measuring from the center line of one dividing partition to the center line of the other.

Frequently, usable area is calculated from the glass line of exterior walls. Again, this is inconsistent with the above definition, as sometimes HVAC units placed along exterior walls protrude into the space up to as much as 24 inches. There are buildings, especially all-concrete buildings, where exterior columns are quite large and consume space. Finally, what about interior columns? These are usually not excluded from the calculations of usable space. The availability of space to satisfy your need for usable area depends upon the efficiency of the building.

The usable area is best calculated by measuring from the glass line of the exterior walls and excludes such spaces as elevator shafts, machine rooms, public stairs, toilet rooms, electrical and telephone closets, mechanical equipment rooms and shafts, and chases penetrating the floors.

According to BOMA (Building Owners and Managers Association International):

> The Usable area of an office shall be computed by measuring to the finished surface of the office side of corridor and other permanent walls, to the center of partitions that separate the office from adjoining Usable Areas, and to the inside finished surface of the dominant portion of the permanent outer building walls.
>
> No deductions shall be made for columns and projections necessary to the building. The Usable Area of a floor shall be equal to the sum of all Usable Areas on that floor.[1]

Rentable Area

Many landlords usually work with a rather nebulous rentable area which, frequently, has no relationship whatsoever to the actual usable area. This rentable area represents the area in square feet written into a lease and is used to calculate the cost of the space. However, you cannot occupy all the space represented by this number. The difference between usable and rentable areas can be quite significant, and unless you have

[1] Reprinted by permission of BOMA International.

measured the usable area for comparison with the rentable area, the difference may surprise you.

The landlord has an investment in the building and somehow he must recover it and show a profit. How this is accomplished depends upon the technique the landlord uses. The landlord has another obligation. In order to make the property attractive to tenants, he must remain competitive in the marketplace. The landlord must fill the building and also show a profit, and he will use whatever technique best suits his purposes.

Your goal must be to obtain the necessary space in the specific location that best satisfies your needs. While on the surface this appears as a battle of wits, it is actually an exercise in futility unless you know exactly how to compare rental costs.

Again, according to BOMA:

> The Rentable Area of a floor shall be computed by measuring to the inside finished surface of the dominant portion of the permanent outer building walls, excluding any major vertical penetrations of the floor.
>
> No deductions shall be made for columns and projections necessary to the building.
>
> The Rentable Area of an office on the floor shall be computed by multiplying the Usable Area of that office by the quotient of the division of the Rentable Area of the floor by the Usable Area of the floor resulting in the R/U Ratio described below.

CONVERSION FORMULA

$$\frac{\text{Rentable Area}}{\text{Usable Area}} = \text{Rentable/Usable Ratio (R/U Ratio)}$$

$$\text{Usable Area} \times \text{R/U Ratio} = \text{Rentable Area}$$

$$\frac{\text{Rentable Area}}{\text{R/U Ratio}} = \text{Usable Area}$$ [1]

The relationship between rentable area of a floor and the usable area of a floor varies from building to building, and from city to city, making this comparison an almost imponderable situation. Therefore, you should remember that your primary concern is the amount of space you use and the amount of money that space costs. No other numbers or figures in a lease have any significance whatsoever in comparing rental costs.

The method of measurement of space often depends upon local custom. New York, for example, has historically had a substantially different method of measuring office space and, therefore, in quoting rent per square foot. There is no legal requirement for an office building or developer to use any standard method. Therefore, it is important for the tenant to clearly understand the method of space measurement used in dealing with the unit cost of space and, if different methods are presented, to convert all figures to a comparable basis for evaluation purposes.

BUILDING EFFICIENCY

An important consideration when trying to evaluate or compare space is building efficiency. This is frequently less important when leasing part of a floor only as compared to leasing an entire building. Suppose you have determined that you need 100,000 square feet of space and you read an ad that offers a building with 100,000 square feet. This building will very likely not satisfy your requirements. How can this be, you ask? The space advertised is what is known as "gross" or "construction" area. It is usually calculated by the architect and it includes the entire area of the building.

Since the construction of a building includes all areas where work is done, it includes the exterior walls. Therefore, the measurement is from the outside of exterior walls and includes all interior spaces such as elevator shafts, stairs, pipe and duct shafts, mechanical and electrical spaces, toilet rooms and similar areas. It is readily apparent that much of what is included is indeed not usable. Even the type of construction may have a significant impact on building efficiency. For example, an all-concrete building with massive exterior walls will yield a lower efficiency than a steel building with steel curtain walls. Of course, a single-story building with packaged rooftop HVAC units will have the highest building efficiency.

In the final analysis of comparing rental costs, building efficiency will have no effect whatsoever, in that you are comparing usable space with usable space only, and all other factors are usually discarded. However, when comparing similar type buildings similarly constructed, you may find it easier and simpler to compare gross area. In these cases, you should be aware of what is included in whole building advertisements and how to use it to your advantage.

FIXED COST

A landlord looks at a building as an investment no different from other investments. His concern is the bottom line or, more accurately, the return on investment. Whether the landlord uses his own money for the construction or borrows from a lending institution, his return on the investment must be sufficient to cover this liability plus some surplus as compensation for his troubles.

In simple terms, the landlord borrows the cash required for the building project, and has a fixed term, fixed rate mortgage. He must satisfy this obligation on a regular monthly or annual basis for the life of the mortgage. This constitutes the landlord's fixed cost. If a tenant leases the entire building and separately pays all other costs such as taxes, fuels, utilities, maintenance, the landlord could base the rental charges on this one cost. Since this is almost never the case, he is faced with variable costs that he usually has no control over, and that must be passed on in some equitable and profitable manner to the tenant.

In years past, fixed costs received serious consideration and variable costs were frequently included as a fixed amount with some form of escalation to compensate for increases in operating costs.

VARIABLE COSTS

Variable costs are, as the name implies, variable, and these costs do vary with time and always move upward. Back in the 1950s when inflation averaged about 1.5 percent annually, rent escalation clauses were infrequently imposed. Leases were then usually written for a short term to provide the landlord reasonable security on his investment, and simplified the problems of a tenant when comparing rental costs. As long-term leases became more common, and as inflation and operating expenses continued to escalate, the landlords need a mechanism to protect the purchasing power of the portion of the rent associated with profit.

Variable costs, in the context of this chapter, are those associated with a leasing situation that do not remain constant. These include such items as real estate tax, operating expenses, and utility and fuel costs. The major problem encountered by tenants is not with initial rental charges, but in miscalculating the future impact of the variable costs. These costs are passed on to a tenant through escalation clauses, and could generate charges that may, in the later years of a lease, completely distort the evaluation of initial rental costs.

A lease, to be profitable to a landlord, must be written in such a manner as to attempt to provide a constant profit value, and not constant dollars. Over the years many different methods have been used by landlords to attempt to compensate for these escalating costs with varying degrees of success. One increasingly popular method of achieving this has been to increase the annual rental by the percentage increase in the Consumer Price Index (CPI). Indexed escalation is often applied to all or a portion of the rental. This practice is so varied that no attempt will be made to cover the many possible combinations, suffice it to alert you to the practice.

COMPARING RENTAL COSTS

Merely comparing the cost in dollars per square foot of one building against the others is not adequate because there are entirely too many variables involved in any leasing situation. First of all, there is no simple uniform method for determining the true cost of the space. The true cost is that which a tenant pays for space he uses, and this usually consists of the following:

1. Rental cost based on dollars per square foot of usable space
2. Facilitation costs
3. Maintenance and operating costs
4. Taxes
5. Utilities' cost

All of your economical comparison problems would disappear if—and that's a big if—all landlords were to charge only for actual usable space, and all facilitation, maintenance, and utilities costs were included. Many landlords usually work with a nebulous "rentable" area which frequently has no relationship whatsoever to actual "usable" area. Landlords usually develop their own unique building standard "workletter." This workletter describes what that landlord agrees to do or provide at his expense in building standard manner and material as part of rent costs. Finally, some landlords charge extra for cleaning, maintenance, and/or utilities.

There is no simple method by which these various costs can be combined and factored to provide a uniform system for economical evaluation. However, by taking these various items singly, a foolproof method can be developed that will provide the necessary answers to your problem of cost comparison.

First, and most importantly, you must determine the actual cost in dollars per square foot for the space you are using; note, *using* and not *renting*. These are two entirely different numbers of square feet for the same space. The rentable area is the area in square feet that is written into the lease and is used to calculate the cost of the space. However, you cannot occupy all the space represented by this number.

What about the area that you can occupy? This is called "usable" area and is defined as that area in square feet that can be measured within the demised premises, measured from the inside of the exterior walls or air-conditioning units if they protrude into the room and excludes such "nonusable" (common use) areas as the central core which houses elevators, stairs, toilet rooms, elevator lobbies, mechanical and electrical rooms, pipe and duct shafts, etc., or in other words, anywhere you can install carpet and place furniture. The difference between usable and rentable area is quite significant, and unless you have taken the time to measure the usable area for comparison with the rentable area, the difference will surprise you.

Table 3-1 Rentable/Usable Area Cost Comparison was compiled for four buildings in a medium-sized Eastern city and the results are startling. Take Building 1 for instance, with its cost of $10.00 per square foot of rentable space. In comparison to the others, this appears to be by far the most expensive; but when compared on a usable area basis, it is the most economical space of the four buildings listed. And of course, the opposite is true for Building 4 with its low cost of $8.66 per square foot of rentable space. There is a valuable lesson in Table 3-1. On your next assignment, make your own table, conduct your own comparison, and you may be surprised at what you find.

Just follow the simple procedure outlined below and you can make your own comparison table:

1. Obtain from the landlord the rentable area in square feet.
2. Obtain from the landlord the rentable area cost in dollars per square foot.
3. Obtain from the landlord an architectural drawing of the area, floor, or floors under consideration.
4. Measure the usable floor area as defined above.
5. Use the formula included in Table 3-1 and calculate the usable area cost.
6. Construct your own cost comparison table.

There is another term introduced in Table 3-1: "Gross" area in square feet. This represents all the space in the area or on the floor as may be the case and is obtained by

Table 3-1. Rentable/Usable Area Cost Comparison.

	AREA IN SQUARE FEET			RENT COST IN $/SQ. FT.	
LOCATION	RENTABLE	GROSS	USABLE	RENTABLE	USABLE
Building 1	17,700	17,700	16,715	$10.00	$10.59
2	24,875	24,300	20,634	8.80	10.61
3	34,200	32,830	28,020	8.75	10.68
4	31,200	30,050	25,018	8.66	10.80

$$\text{Usable Area Cost} = \frac{\text{Rentable Area}}{\text{Usable Area}} \times \text{Rentable Area Cost}$$

measuring from the exterior glass lines in buildings consisting mainly of glass exteriors (curtain wall construction); or from the exterior of the exterior brick or block walls where this solid wall predominates. Where interior partitions bound the area, measure from the centerline of these partitions. In all cases, the gross space includes everything within these boundaries as shown in Figure 3-1 and for that building.

Gross area = length × width

140 × 185 = 25,900 square feet

While the area so calculated serves no useful purpose in a leasing situation, it is used to determine the efficiency of design of a particular space as follows:

$$\% \text{ efficiency} = \frac{\text{Usable area}}{\text{Gross area}} \times 100$$

and for the building shown, the efficiency is

$$\frac{21{,}100}{25{,}900} \times 100 = 81\%$$

The next most significant consideration in a leasing situation is the landlord's "Workletter." This describes the work that the landlord will provide at his cost and details what the tenant will have to pay on a per unit basis when he exceeds that allowed by the landlord as part of the rent cost. Three sample workletters are included and are typical of the types encountered in leasing situations. The quantities described under landlord's work and the related costs listed under tenant's extra costs are not intended as firm quantities and costs to be used in lease negotiations. They are included as encountered in past situations and could serve as guides if used judiciously. These workletters are attached to and become part of the lease agreement.

Sample No. 1 is the simplest type in use and clearly defines responsibilities and costs. For instance, the landlord will provide, as part of the rent cost, 8 linear feet of interior partitioning for each 100 square feet of space leased and will charge $22 for each additional linear foot of partitioning needed to satisfy the requirements of the tenant. With a lease that contains this type of workletter, the landlord can quote a firm rental cost because he knows exactly what his financial obligation is for any amount of space, small or large. The tenant can also calculate his additional costs as soon as his drawings are completed.

These tenant extras' costs are usually payable in cash as soon as the work is finished, but some landlords would be willing to include these costs as additional rent prorated over the life of the lease.

Sample No. 2 is somewhat more complex and is usually developed after advanced negotiations between the landlord and the tenant. It is more detailed and covers specific items reflecting input by a tenant who has clearly defined requirements based upon facilitation standards for his company.

This workletter, in addition to defining the landlord's responsibility, provides allowances for such amenities as wood paneling, vinyl wall covering, carpeting, drapery. It also allows for extras and credits as may be the case. This additional detail permits more latitude in the design of the space after the lease cost has been negotiated.

Sample No. 3 is a modified workletter in that it usually consists of a paragraph in the lease proper under a section usually identified as "Improvements" and refers to an attachment, Schedule A. This Schedule A is developed to show what the landlord will do at his cost and, in this case, includes all necessary facilitation to satisfy the tenant's needs.

SAMPLE NO. 1—WORKLETTER
LANDLORD BUILDING STANDARD WORK

The landlord agrees to perform at his expense, the following work (landlord's work) in building standard manner and materials in the office space only of the demised premises.

LANDWORK'S WORK	TENANT EXTRAS COST
Corridor Doorway 1 each tenant	$400.00 each
Interior Doorway 1 each 300 sq. ft.	250.00 each
Interior Partitioning 8 linear ft. each 100 sq. ft.	22.00 per linear ft.
Lighting Fixture 1 each 80 sq. ft.	110.00 each
Duplex Convenience Outlets 1 each 100 sq. ft.	60.00 each
Clock Outlets	65.00 each
Convenience Outlets (Floor)	95.00 each
Telephone Outlets (wall)	45.00 each
Telephone Outlets (floor)	95.00 each
Vinyl Wall covering	1.25 per sq ft of wall area

SAMPLE NO. 2—WORKLETTER

To induce Tenant to enter into a Lease and in consideration of the mutual covenants hereinafter contained, the Landlord and Tenant mutually agree as follows:

1. (a) Tenant, at Tenant's sole expense, shall cause to be prepared complete, finished, detailed architectural drawings and specifications for Tenant's partition layout, reflected ceiling and other installations for work to be done by Landlord under Paragraph 2 hereof.
   (b) Landlord agrees to pay for electrical and mechanical drawings and specifications for work to be done under Paragraph 2 hereof.
2. Landlord agrees, at its sole expense and without charge to Tenant, to do the following work in the demised premises:

(a) Supply and install Landlord's building standard partitions, located wherever shown on said plans described in Paragraph 1 (a).
(b) Supply and install Landlord's building standard doors, bucks, and hardware consisting of office door latch sets, locksets, and hinges wherever shown on said plans described in Paragraph 1 (a).
(c) Supply and install Landlord's building standard suspended fissured mineral ceiling tile system with concealed spline throughout.
(d) Supply and install Landlord's building standard venetian blinds in all peripheral windows.
(e) Supply Landlord's building standard electrical facilities sufficient for a connected load of $3\frac{1}{2}$ watts at 480/277 volts, three phase, and $1\frac{1}{2}$ watts at 208/120 volts, three phase, per square foot of each area served.
(f) Supply and install Landlord's building standard recessed fluorescent lighting fixtures in the amount necessary to provide average maintained lighting of 80 foot candles at desk height.
(g) Supply and install Landlord's building standard air-conditioning system with necessary ductwork, to serve interior areas. Supply and install Landlord's building standard perimeter induction system with thermostats in an amount not exceeding one (1) thermostat for every five (5) peripheral window units. Said system shall be designed to be capable of maintaining, within tolerances normal in first class office buildings and subject to density factors of not more than one (1) person per 100 square feet nor more than 5 watts of electrical load per square foot of each area served, inside space conditions of 78° F dry bulb and 50% relative humidity when outside conditions are 95° F dry bulb and 75° F wet bulb.
(h) Supply and install Landlord's building standard duplex wall receptacles in the amount of one (1) for each 150 square feet of rentable area; and building standard wall switches in the amount of one (1) for each office and as required by code in open areas.
(i) Supply and install Landlord's building standard telephone wall outlets in the amount of one (1) for each 150 square feet of rentable space.
(j) Supply and install Landlord's standard $\frac{1}{8}$th inch grade "C" vinyl asbestos tile floor covering in all uncarpeted areas.
(k) Paint the peripheral walls and all building standard partitions and doors in Landlord's standard manner, colors to be selected by Tenant from Landlord's standard colors, allowing two colors for each room or area, except where paneling or vinyl wall covering is specified.
(l) Supply and install wood paneling located wherever shown on said plans described in Paragraph 1 (a) at a cost not to exceed $3.00 per square foot of wall area as selected and specified by Tenant.
(m) Supply and install vinyl wall covering located wherever shown on said plans described in Paragraph 1 (a) at a cost not to exceed $0.50 per square foot of wall area as selected and specified by Tenant.
(n) Supply and install carpet located wherever shown as said plans described in Paragraph 1 (a) at a cost not to exceed $17.50 per square yard as selected and specified by Tenant.
(o) Supply and install drapery located wherever shown on said plans described in Paragraph 1 (a) at a cost not to exceed $20.00 per lineal foot of length of window wall as selected and specified by Tenant.

3. Building standard unit prices; extras and credits.

ITEM	EXTRA	CREDIT
a. Light fixtures	$ 75.00/ea.	$55.00/ea.
b. Floor receptacles	60.00/ea.	—
c. Telephone floor outlets	55.00/ea.	—
d. Perimeter A/C		
1. Thermostats	90.00/ea.	—
2. Valves	220.00/ea.	—
e. "C" grade floor tile	—	.20/sq. ft.
f. Building standard paint	—	.10/sq. ft.

SAMPLE NO. 3—WORKLETTER

IMPROVEMENTS: (Paragraph in lease agreement)

1. The parties have signed outline specifications identified as Schedule A and described and shown on Drawings No. 74-0604-1 through 6, copies of which are attached hereto, and made part of this lease, describing improvements to be provided and installed by Landlord, at Landlord's expense, and the Landlord agrees to perform the work and make the installations in the leased premises which are set forth in Schedule A and Drawings No. 74-0604-1 through 6.

SCHEDULE A—Description of Landlord's Work

1. Doors from public corridor to demised premises will be solid core $1\frac{3}{4}''$ × 3′—0″ × 7′—0″ flush plain-sliced red oak door with mail slot and equipped with lockset, lever door handle, and door closer.
2. Partitions within tenant's area will be $2\frac{1}{2}''$ steel stud with $\frac{1}{2}''$ gypsum board finished to the hung ceiling line, taped, spackled, and painted.
3. Ceilings will be suspended acoustical tile on exposed 2′ × 4′ grid system.
4. Painting throughout demised premises will be two coats as selected by tenant from standard colors provided by Landlord, and not to exceed two (2) colors per room or area.
5. Heating, ventilating and cooling will be provided throughout the year on business days Monday through Friday, excluding holiday, from 8 A.M. to 6:00 P.M. to maintain temperatures within the demised premises of: not less than 72°F, nor more than 78°F during the heating season, with outdoor temperature range of 0°F to 65°F; 78°F average temperature, and 50% relative humidity during the cooling season when outside temperatures are 95°F dry bulb and 75° wet bulb. The above is based on occupancy not to exceed one person per 100 square feet of rentable floor area, and a maximum electrical load for lighting and/or office machines of 5 watts per square foot of rentable floor area.
6. Lighting will be building standard recessed fluorescent light fixtures with 4-40 watts lamps each unit.
7. All work shown and described on the following drawings.

   No. 76-0604-1—Architectural plan
   2—Partition plan
   3—Reflected ceiling plan

4—Electrical plan
5—Telephone plan
6—Finish schedule

The cost figures included in the sample workletters are not intended for estimating purposes as these costs will vary from city to city throughout the country. These examples should serve as a guide to show the various types used by landlords.

Finally, a few words about maintenance, operation, taxes, and utilities costs. In most large office buildings, these items are included in the rent. In fact, many landlords will not permit anyone other than their own crews to handle cleaning and maintenance work. However, do not overlook these items. They should be included in the lease; if not, ask about them.

Now, back to comparing rental costs. Remember the old cliché, you must "compare apples to apples?" All costs must be included to make the comparison or similar expense items must be deleted from each. For example, some landlords will include leasehold improvements in the lease and amortize it over the life of the lease while others may require that these costs be paid up-front. Once you have determined that your rental costs are on a similar base and represent like conditions, you can make your own comparison as done in Table 3-1.

SUMMARY

The reason to conduct a rental cost comparison is to determine which property is the most economical over the life of the lease, providing, of course, that all conditions of the original criteria are satisfied. The example in Table 3-1 shows a difference of $0.21 between the lowest and highest cost when compared on a usable area basis and will yield a $10,500. savings on a 10,000-square-foot space over a five-year lease. This should be sufficient incentive for anyone to conduct his own rental cost comparison.

A few simple steps should permit you to obtain space at the lowest cost, consistent with your needs:

1. Review the terms and conditions of the lease to insure that all similar costs are included in each.
2. Determine the usable space in each of the buildings under consideration.
3. Calculate the usable area costs.
4. Develop a cost comparison table.
5. Calculate the cost savings over the life of the lease.
   At this point it is essential to carefully check the variable costs as they will surely impact rental costs during the term of the lease. If these are similar in all the leases under evaluation, they can be excluded; otherwise, they must be considered in the economical comparison.
6. Try to convince the president of your company that the space he has arbitrarily selected is not the most economical.
7. Good luck to you in your new place.

4. Consulting and Contracting Services

Before the appropriate consultants and contractors can be utilized effectively, their services must be understood fully. Unfortunately, all too many facilities planners and managers do not understand the advantages of using consulting services. This is particularly true of companies with a small facilities staff. At times, even large companies with an adequate professional facilities staff do not fully utilize these services to best advantages. There are several reasons for this. Possibly the single most important reason why consulting and contracting services are not used, is because of a lack of understanding of the range of services offered. In addition, many facilities managers feel that when they perform these services, they can better interpret their own needs and have better control over the entire project. Finally, these services are not more fully utilized because many facilities planners and managers do not know at just what stage of the project the services should be retained.

The type of service required should match the project under consideration. This could range all the way from real estate specialists to assist with site search, through an architectural/engineering firm for the design construction project, to a general contracting company for the actual construction work. The choice is difficult because there is no clear-cut division between the scope of services offered by the various consulting firms and contracting companies. In fact, there are firms that will perform every service from site search through building design and construction. All you need do is explain the intent of your project and the firm will place you in a building at some future date at a guaranteed maximum cost. Or you can deal with a number of firms, with each performing a specific task assigned by you. There is no one best method for all projects.

Many companies and corporations are involved infrequently in major projects, and some only once during the lifetime of their key executives. Some corporations have fully staffed facilities departments while others have no one professionally qualified to do this work. From this it is obvious that the problem could be enormous for some facilities managers, and rather simple for others. But since the companies that lack appropriate in-house talent far exceed those that are fully staffed, the safest approach is to discuss the range of services and types of firms available. This is best accomplished initially by separating the consulting firm from the contracting company, even though later it will be shown that there is frequently significant overlap of services offered.

CONSULTING SERVICES

This section will deal with real estate development firms, architectural and engineering firms. The next section will address general contractors, subcontractors and construction managers.

Facility planners can frequently obtain free services from state development commissions, utility companies, chambers of commerce, etc. These services are of great assistance to the newcomer, as well as the old pro. While this is usually a good way to begin site selection and relocation planning, additional services are frequently required, particularly where new facility construction is included in the project. This is where the specialist, the independent consulting firm, with its staff of architects, engineers, planners, developers, and economists can best satisfy the needs of corporate planners.

Real Estate Consulting

These firms specialize in site selection and economic evaluation. This could be an essential first step in any major relocation project. The method of compensation is usually on a fixed fee basis, roughly equal to a real estate broker's commission. The difference between the two is that the fee is paid whether the deal is consummated or not. This is similar to compensation for any consulting service. The actual fee could be a fixed fee based on the scope of the project or could be direct labor costs plus overhead and profit, and out-of-pocket expenses. The latter could also have a guaranteed maximum that will limit the total liability and is usually billed on a monthly basis listing all employees involved with the agreed upon markup.

The services required should be clearly defined and can, in some cases, be contracted for in several phases. The separate phase method could be an excellent hedge, particularly where the project is not fully defined and absolutely committed.

These firms with their specialized expertise and experience can analyze your requirements, investigate and evaluate a number of sites, and recommend the site best suited to your needs.

There are as many major reasons for constructing new facilities as there are reasons for selecting a particular site. Before you can explain to others just what your plans are, you must be able to define the reason for constructing the new facility and the reason for selecting the new site. The Facility Planners' Checklist shows the results of several surveys conducted over the past two decades, listing the reasons for constructing new facilities and selecting new sites in order of importance most frequently selected. With this table, facility planners can establish their own criteria by assigning percentage values to the various items, i.e., 100 percent most important, 0 percent least important. Since individual site selection factors emerge, grow, and then recede in importance, or the complexity of a particular factor changes over the years, it is essential to approach each new situation in the light of current conditions. While this list is fairly comprehensive, you may have reasons not listed that should be included. While this checklist is primarily designed for industrial plants, it will serve equally as well for offices. Unfortunately, offices, and particularly corporate headquarters buildings, are usually sited according to the chief executive's preference.

FACILITY PLANNERS' CHECKLIST

Major Reasons for Constructing New Facility	*Check one*
Expand production	________
Consolidate production	________
Relocate to a new area	________
Replace existing obsolete facility	________
Build in a new market	________
Adjacency to raw materials	________
For centralized distribution	________
________	________
________	________

Major Reasons for Selecting New Site	*Apply Percentage*
Availability of right type of labor	________
Favorable labor rates	________
Raw material availability	________
Utilities availability	________
Utilities cost advantage	________
Market center	________
Tax advantage	________
Land cost advantage	________
New construction financing advantage	________
Freight transportation services	________
Community acceptability	________
Favorable climate	________
Educational facilities	________
Recreational facilities	________
________	________

New Plant Site Setting	*Check one*
Urban	________
Suburban	________
Rural	________

The type and size of firm considered should be consistant with the needs of the project. While this may sound like an oversimplification, it is certainly not meant to be. You must first analyze the scope of the project before you can be in a position to seek professional assistance. Let us suppose that the project consists of a small office space in a particular city. You may feel that you are fully capable of handling all aspects of site search, lease negotiations, rental cost comparison, space planning and design, and you may well be. In that case, you will find elsewhere in this book all the necessary information to lead you through the entire process. Suppose that you are inexperienced in space planning, yet plan to do it yourself. You may well end up with inefficient space utilization. In this case, you will be paying rent for space that you do not need. A professional space planner may be justified in even this simple example.

Now consider a large project that includes a new manufacturing facility that must be adjacent to a raw material source and centrally located with respect to the finished product market. In this case, there may be several sites in several locations that may appear to equally satisfy the above criteria. In a project of this complexity, even the selection of a specific site may have significant economic import. The money misspent on the wrong site or location may be the difference between profit and loss.

After you have determined that you require the services of a real estate consulting firm, you must analyze what they offer to compare that to your needs. A fully staffed firm will provide some or all of the following:

- Assist in evaluating the project needs
- Assist in fully investigating, analyzing, and recommending potential sites
- Assist in negotiating for, and acquiring, the final chosen site
- Assist in obtaining all necessary and appropriate approvals, permits, and consents needed for the project's construction
- Insure that the project has all public and private utilities and other services needed to meet the project's objectives.
- Assist in the selection of other consulting services needed to carry the project to completion

As you will see later, some of these services overlap those offered by other consulting firms. That is not necessarily of major significance because the services of the various consulting firms can be coordinated to eliminate the unnecessary functions.

The major decision you will be called upon to make is to determine what type of consulting services you should retain first. That depends to a large extent upon the project at hand. You must determine what is most important to the project. Several such elements are listed below as a guide, simply indicate by percentage after each to determine the values attached to the individual items.

Item	*Percentage*
Site	__________
Location	__________
Building	__________
Economics	__________
Image	__________

For those inexperienced in dealing with consulting firms, a few words of advice:

It usually does not cost anything to talk to a member of a firm. In most cases, representatives will be glad to come to your office to discuss what the firm offers and how it can help you.

Most firms offer a broad range of separate services starting with simple studies and recommendations at some minimum fixed fee.

Many firms offer services that can be tailored to the specific needs of the project.

Some firms will offer their specialists who will work with your staff to develop a project scope. These services are usually offered on an as-needed, per diem basis.

If there is any doubt in your mind as to what to do about consulting services, the safest route is to retain a professional, if for no other reason than to discuss the project and his probable involvement. At worst, you will have spent a little money in exchange for peace of mind.

Architectural Services

While some architects provide only architectural services, many are A/E (Architectural/Engineering) Firms and offer a full range of services that may include all, but may not be limited to, the following:

Architectural
Space planning
Interior designer
Interior decorator
Architectural acousters
Lighting consultant
Food service consultant
Elevator consultant
Site planner
Electrical engineering
Mechanical engineering
Structural engineering
Civil engineering
Fire protection and life safety
Lightning protection
Soils consultant
Testing services

There are several advantages to using full-service firms. First and foremost, you will deal with only one person, and that person will have total responsibility for your project. This greatly simplifies your involvement, and permits you to accomplish your work without having to devote all your time, and maybe that of several subordinates, to the project.

Further, it is the responsibility of the firm to provide a fully coordinated design that includes all considerations on the part of the specialists involved. Finally, in the event of design coordination, the responsibility lies with one firm.

Of course, there are disadvantages associated with the use of full-service firms. You will have to pay for the special effort expended by the firm for the control and coordination of its professional specialists. Unless you have a staff capable of dealing with the various specialists, the full-service firm is your best choice.

You may not have any voice in the selection of the specific specialists the firm uses. This may or may not present problems, depending upon the circumstances. In those cases where you have a strong desire to use your own specialists, e.g., for food service, interior designer, site planning, you well may be able to convince the architect to include them in the contract.

Your second choice is the use of individual professional specialists. These are individuals or a group of individuals working together and offering services limited to a

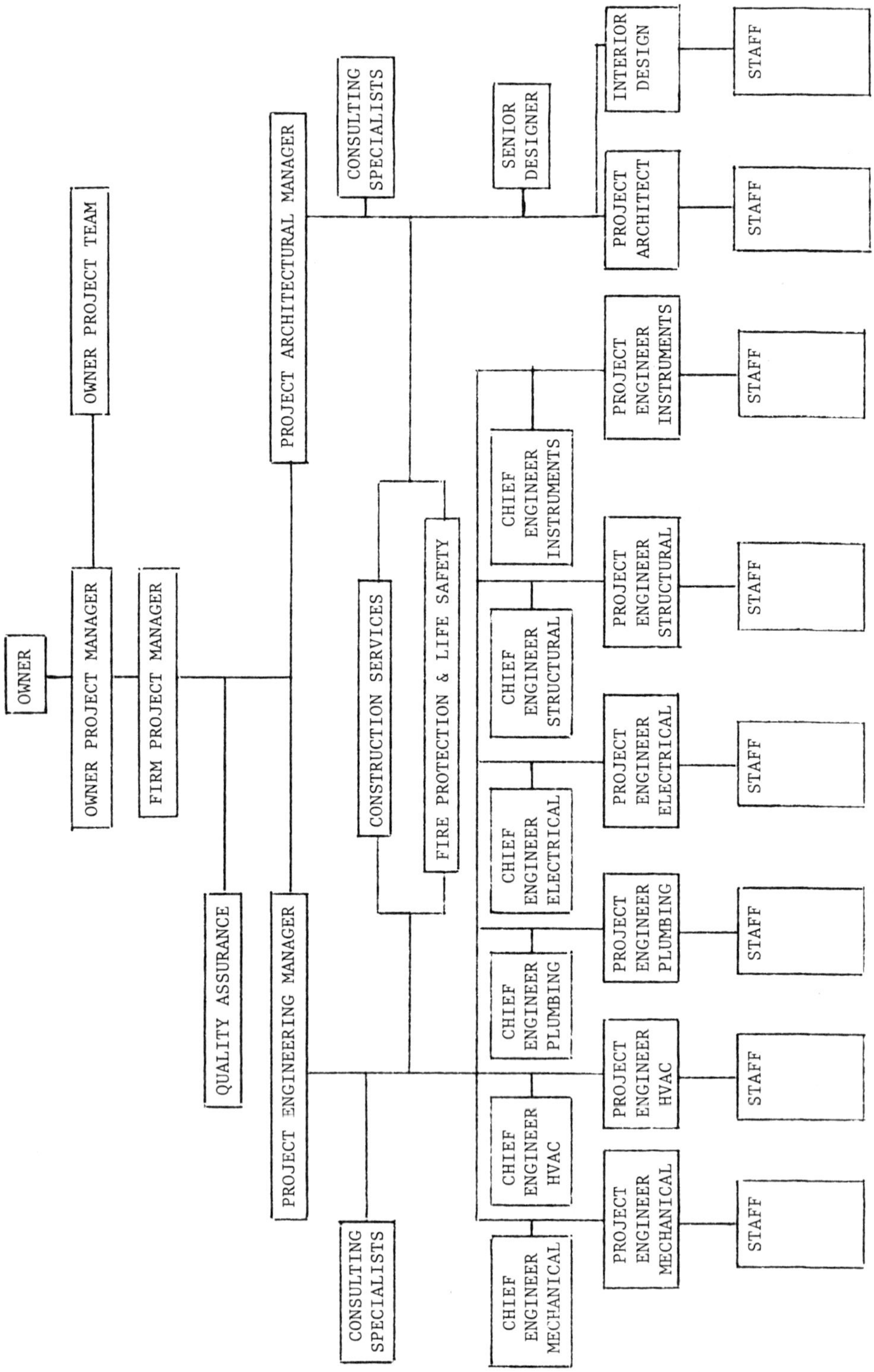

Figure 4-1. Project Team—Full Service Concept.

narrow field or specialty. Profession specialists are experts and practice one particular line of work. The range of specialists runs the full gamut from soils testing to solar energy and includes almost everything in between. The major disadvantage with using individual specialists is that you may not know of everyone that is required and may inadvertently omit a critical specialist.

From this it is obvious that your choice also depends upon your anticipated degree of participation and your qualifications and experience with the various specialists. Unless you feel adequately qualified to handle the myriad of problems usually involved in dealing with separate specialists, your best choice is the full-service firm.

Engineering Services

There is usually another possible combination of services available to the facilities manager, and this is the separate engineering firm. In years past, the architect was

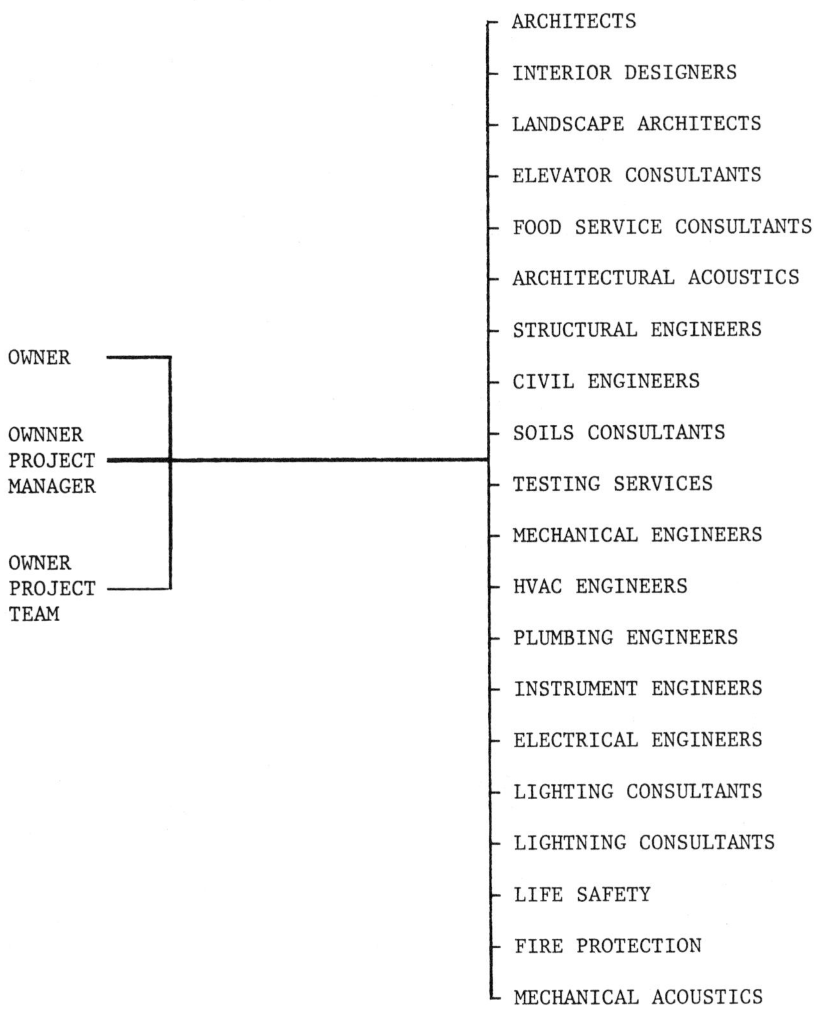

Figure 4-2. Project Team—Individual Specialist Concept.

considered the prime consultant, but that has been slowly changing. This is particularly true where the project is primarily an engineered facility, i.e., a manufacturing plant, processing plant, power plant, etc., where the engineer becomes the prime. There are also instances where facilities managers use separate architectural and engineering firms with both reporting to the owner.

Again, in years past, most buildings were primarily architectural in nature, and other than structural engineering, electrical and mechanical engineering for the buildings were very simple and straightforward. That has all changed, particularly with the advent of highly automated mechanical systems. This has evolved to such a highly complex state that many engineering firms utilize a separate discipline group to design system automation. Electrical systems are also becoming more sophisticated, particularly with the introduction of complex fire alarms, building security, and entrance and exit monitoring and control. It is for these reasons that there are more firms offering only engineering services.

Today, facilities managers are more frequently contracting with separate architectural and engineering firms to satisfy their highly specialized needs. With this approach, you can select an architect whose reputation and design experience ideally suits the current project criteria, and then pick an engineer with the appropriate technical expertise. In this manner you can have the best of both worlds.

Recently, architectural and engineering firms whose design concepts are compatible are offering joint services to clients. Some architects have aligned themselves with several engineers to offer a wider range of services, and engineers have done the same. This permits both firms to submit design proposals for a wider range of projects as the combined services can be better tailored to the client's needs.

ORGANIZATION CHART OF SEPARATE CONSULTANTS

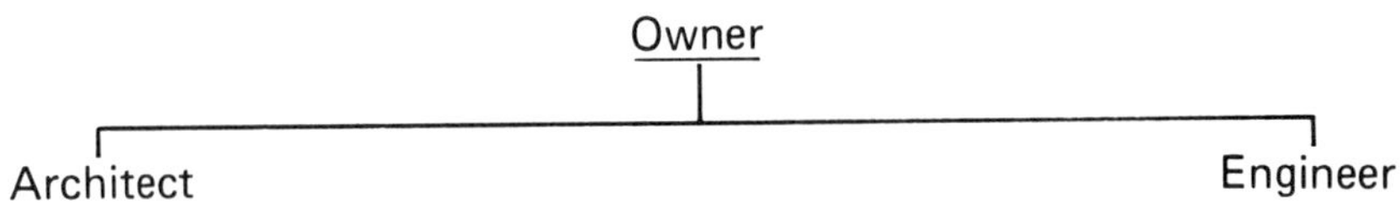

CONTRACTING SERVICES

The purist may wish to take exception to the heading of this section and may prefer construction contracting as a title, but that does not address the following subject matter fully either.

Contracting services, similar to consulting services in several basic types, are of a number of different variations. There are four types commonly used:

Design/build contracting
Construction manager with multiple separate-trade contracting

Single prime (general) contracting
Multiple separate-trade contracting

All four are available as guaranteed maximum construction contracts (GMP) with a percentage of savings accruing to the contractor, but the last two are usually bid at a fixed price. In any case, the facilities manager must decide at the beginning of the construction project which method of contracting he intends to use because the A/E firm must develop drawings and specifications to suit the method selected. Each requires a set of contract documents; drawings and specifications, that may be different.

Each method has its advantages and disadvantages. Privately financed projects most frequently use general contracting with a fixed price bid. Public works projects require multiple separate-trade contracts with fixed price bid. Where time is of the essence, privately financed projects use a design/build firm, particularly where the project is unique and the design/build firm has expertise in the particular project. The same time advantage can be obtained using an A/E firm and *fast-track* construction.

Fast-tracking is accomplished by designing and bidding individual construction contracts matching construction concepts. For example, in an all-concrete building, the site work is completed first and bid immediately. While this is being done, the substructure, footings, and foundations are designed and bid. While this work is being done, the superstructure is designed and bid. Lastly, the electrical and mechanical portion of the project is designed and bid. This usually places an additional burden on the A/E firm and increases the design cost of the project. This added cost is usually recovered many fold by the time saved from start of design to actual occupancy. The fast-track method can save anywhere from three to six months on the entire project.

Design/Build Contracting

There are several large firms throughout the country that specialize in this type of contracting. The primary advantage to this technique is that one firm is responsible for the entire project and frequently the project can be completed in the shortest span of time. Unfortunately, this is usually not the most economical choice, and frequently the owner has no control over much of what goes into the building. This method is most frequently used for manufacturing and processing plants, where delays in construction completion time affects on-line production, and results in loss of revenue. You will have to decide whether or not this method serves your interests to the greatest extent.

ORGANIZATION CHART OF DESIGN/BUILD CONSULTANT

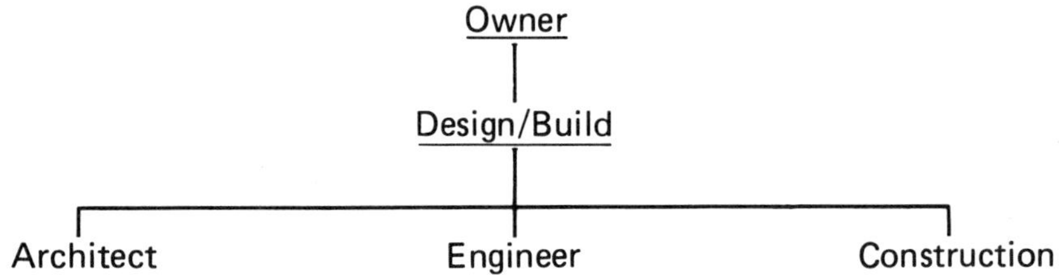

Construction Management with Multiple Separate-trade Contracting

Construction management is practiced in so many different ways that it is difficult to define clearly what constitutes a construction manager. Some general contractors offer this service to manage their own group or they will manage independent contractors. Some firms offer construction management services and have no contracting group. Some architects and engineers offer this service and will manage the construction phase of the project. As you can see, this technique could be replete with problems and there could also be conflicts of interest. Reputable construction managers could be a definite asset to owners. They are street wise and know a great deal about the construction industry. Basically, they are business managers, coordinators, and negotiators.

When considering a construction manager, it is essential that you check his track record. What has the individual done lately on projects similar to yours? This should be the sole criterion for the inclusion of a construction manager in your plans.

Single Prime (General) Contracting

This is possibly the simplest and generally the safest method contracting service. The A/E firm designs the building and the general contractor agrees to build it for a bid fixed fee. At the outset of the project, you know what the design services will cost and after selected contractors bid for the construction work, you know what the completed project will cost. Tack onto this a 5 to 10 percent contingency for unforeseen problems and you are home free. This contingency is necessary because there has never been a perfect set of contract documents for a building that is to be constructed as designed. Secondly, there are conditions beyond the control of the contractor that adversely affect construction. Thirdly, vendors have been known to go out of business between the time the design is started and construction is in progress. Then there is the ever-present concern of strikes. These could delay critical aspects of construction into unfavorable periods of weather.

As with design firms, select contractors with a track record in the type and size of your project. Normally, contractors who do work in the $2 million range should not be included in a $30 million project. Similarly, contractors that specialize in office renovation should not be considered for a major processing plant project. What this really means is that contractors should be prequalified. Prequalification involves evaluating the list of contractors who have expressed an interest in the project. Ask each to complete a questionnaire that furnishes information about the firm's project experience, financial capability, and stability.

After all bids have been received, analyze them carefully. Beware of the contractor who is unreasonably low because most likely something was missed and holding the contractor to that low bid will not serve the project. Driving a contractor into bankruptcy will delay the project and eventually cost more. You can alert the contractor to its apparent omission if it is considered a reliable contractor, or you may wish to exclude it. The low bid should never be the only consideration when choosing a contractor. Be sure that contractor can complete the project within budget and time.

ORGANIZATION CHART OF GENERAL CONTRACTING

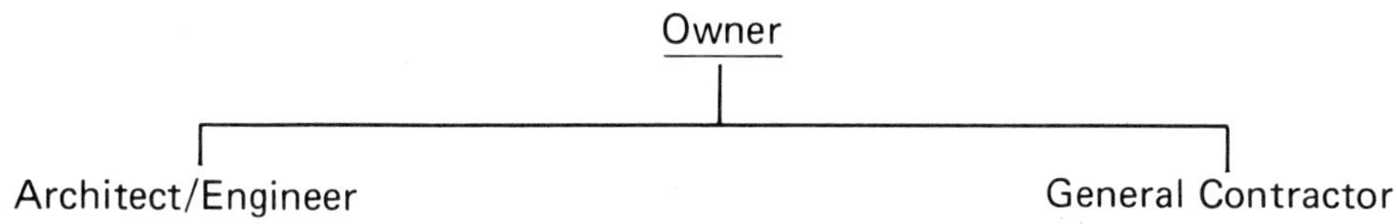

This chart is very significant. Please note that both the A/E firm and the general contractor are responsible to you the owner. You have a contract with both. You must direct the efforts of both.

The American Institute of Architects has developed documents that address, among other subjects:

- Standard Form of Agreement Between Owner and Architect
- Standard Form of Agreement Between Owner and Contractor
- General Conditions of the Contract for Construction

Under Article 2, Architect, 2.1 Definition, 2.1.2. ''Nothing contained in the Contract Document shall create any contractual relationship between the Architect and the Contractor.''

Multiple Separate-Trade Contracting

This is probably the most difficult contract to administer. Most public works projects use this method. It is particularly common in Eastern and Midwestern states having statutes that require municipal projects to be let as separate contracts divided by trade, i.e., general construction, plumbing, HVAC, electrical, etc.

The major disadvantage of separate contracting is that the owner assumes an increased responsibility for the performance of the overall project. The owner essentially assumes the role of construction manager, whether intended or not.

The major advantage to the owner is that the separate contractor eliminates the multiple markup on equipment and materials common in general contracting, and enhances competition among the bidders resulting in lower construction costs.

Unless you have an adequate staff to assume the role of construction manager, this type of contract is not for you. This may be well on smaller projects but could present insurmmountable obstacles on large, complex projects.

ORGANIZATION CHART OF MULTIPLE SEPARATE-TRADE CONTRACTING

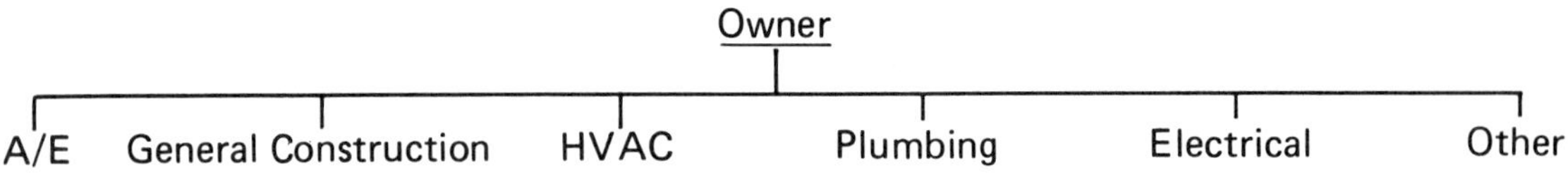

SUMMARY

From this it is obvious that you have many options, and the selection depends upon the size and capability of your staff, and the size and complexity of your project. The success or failure of the project is entirely in your hands—a wrong choice could yield disastrous results. In the event that there are any doubts in your mind, you should select the method that has the best opportunity for success.

Secondly, unless you have a thorough knowledge of the project and the intended end results, you could well be doomed from the beginning. If behoves you to have a clear concept of your objectives. If you have no one on your staff who is capable of overseeing the project, it will be to your advantage to hire someone on a short-term, one-project basis. This is essential where new building construction is a once-in-a-lifetime requirement. There are many experienced professionals who are available and who have necessary expertise that will enable them to handle such work. One word of caution. Make certain that the person selected has the necessary credentials, and then verify it by checking his performance.

A simple step-by-step procedure should be a valuable aid:

1. Evaluate the qualifications of your staff.
2. Develop the scope of the project.
3. Establish an obtainable goal.
4. Determine the type of consulting and design services required.
5. Talk to several firms on an informal basis.
6. Develop a reasonable schedule for the project starting with the selection of the consultant through occupancy of the facility.
7. Develop an adequate budget for the entire project remembering to include for contingencies and escalations.
8. Be reasonable with your demands of both the consultants and contractors. If a consultant or contractor goes into bankruptcy during the life of the project, it could mean disaster for you.
9. Serious errors and omissions may be committed by design professions when owners unduly accelerate the design schedule.
10. When a project schedule must be realistically shortened, there is a technique called "fast tracking." With this scheme, the construction overlaps the design, for example, as soon as the site and underground portion of the design is completed, this work is subcontracted. When the next phase, substructure, is designed, this work is let for construction. Next comes the superstructure, followed by the electrical and

mechanical work. This technique should be utilized only when time is critical and the work, both design and construction, must be performed by experienced professionals with a track record in fast tracking. Construction time can also be reduced by a technique called "early steel bid." In a building with a steel superstructure, the structural steel is bid and awarded as the design for that portion is completed. This could provide a time savings of from 4 to 10 weeks. In order to reduce confusion and to make a more coordinated construction contract, this steel contract should be assigned to the general contractor as soon as one is selected. Design/build contracting can also be utilized to shorten the construction time period.

11. Enter the project with your eyes wide open. Be realistic in your demands and remember that consultants and contractors are in business to make money and they must make money to stay in business.

5. Selecting Design Consultants

One of the more difficult problems in new building construction and existing building renovation is the selection of design consultants, and the degree of difficulty is directly proportional to the technological complexity of the project. The selection process can be simplified if you have a good fundamental knowledge of the services offered by consultants, and know what you need to do to interface with them effectively. The success or failure of any project depends to a great extent upon the quality of communication between the design consultant and you, the client.

It is just as important to match the consultant to your needs. The firm should be of appropriate size, have the necessary capability, and have experience relevant to the project. The firm should be large enough to absorb your project without undue impact upon its workforce. A too-small firm may not be able to adequately staff the project, whereas a too-large firm may not devote appropriate attention to it. In either case, the project will suffer. There are design consultants that will provide an optimum match for your project.

Some companies have facilities departments that are fully staffed with architects and engineers and go through the selection process on a regular basis, they may encounter no difficulty, or at least think so. Most companies do not have a facilities department or even experienced qualified professionals in this area. Frequently, new building construction or existing building renovation is handled by key executives on a part-time basis.

The first step in the selection process is a review of services offered by design consultants. Since there is no set standard of services for all firms, and the scope of these services depends upon the size and complexity of the project and your specific requirements, the following addresses the full range of available services that you may require.

DESIGN CONSULTANT SERVICES

In order to achieve the desired measure of success on any project, it is necessary that you have a thorough understanding of your objective. Then, and only then, will you be in position to evaluate the services offered by consultants and to utilize these services to best advantage. Most consultants, particularly full-service firms, offer a broad range of industrial and commercial services that may begin with site search and leasing assistance, proceed through building and facilities design and engineering, to move coordination and occupancy. This is implemented by preliminary study and evaluation, cost estimating, construction contract documents, competitive bid analysis, contract award assistance, and administration of construction. Most firms offer services that can be tailored to satisfy client requirements with respect to project scope, degree of participation, budgetary limitations, and personnel involvement. Most firms separate these services into two categories; basic services and additional services. These are further subdivided and clearly delineated in order that the owner and consultant understand exactly what will be done under the terms of the agreement. This agreement, signed at the onset of the project, becomes a legal document defining the responsibilities of both parties.

Basic Services

Since site search and lease negotiations have been covered elsewhere in this book, they are excluded here. This section addresses those design services that are necessary for the implementation of new building construction or existing building renovation. While standard agreements between consultant and owner usually list five phases under basic services, there are additional services available, depending on the scope of the project and the desires and needs of the owner. You may contract for these services individually or totally to suit your requirements. The work performed under basic services is usually accomplished in five phases: predesign; schematic design; preliminary design (or design development); construction contract documentation; bidding or negotiation; and construction administration. The fact that *five phases* are stated and then *six* are listed is not a mistake. Many firms may combine predesign and schematics, but they will be addressed separately here.

Predesign Phase. While this phase is usually performed in the shortest period of time, and consumes the smallest portion of the fee, it is as important as any of the others. This is where the design criteria for the project are established. Effective two-way communication is absolutely essential at this stage, because in order to achieve the desired measure of success the consultant must understand the owner's objective. Therefore, initial responsibility lies with the owner who must define the project in terms of *what, where, when,* and estimated *budget.* The owner must delineate his expectations in terms of appearance, efficiency, feasibility, expandability, utilization, and maintenance. He must also address problem areas, such as, floor loading, fire protection and safety, environmental impact, and special requirements for a computer center, uninterruptable power systems, lighting, power requirements, plumbing requirements, operating schedule, and type of construction contract anticipated. This is not intended as an all-inclusive list—there are variations with the type and complexity of the project.

Since many consultants operate on a project team basis, the consultant will select and identify the members of the team. It is usually headed by a project manager who has total project responsibility and is the primary interface person with owner's representative. Under the project manager are the project engineering manager and project architectural manager. See the Figure 4–1, Project Team—Full Service Concept in the preceding chapter for complete details.

OWNER'S CHECKLIST—PREDESIGN PHASE

1. Project location ________________________________
____
____
2. Owner's representative ________________________________
________________ Phone ( ) ________________
3. Project type ________________________________
4. Size of site, acres ________________________________
5. Size of building, floors ________________________________
area, sq. ft. ________________________________
6. Number of employees ________________________________
7. Operating schedule ________________________________
8. Parking requirements, employees ________________________________
visitors ________________________________
9. Problem floor loading ________________________________
____
10. Fire protection and safety ________________________________
____
11. Computer center ________________________________
12. UPS system ________________________________
13. Lighting ________________________________
14. Electrical power requirements ________________________________
15. Plumbing requirements ________________________________
16. Owner's standards ________________________________
17. Construction contract contemplated ________________________________
18. Anticipated completion date ________________________________
19. Estimated budget ________________________________
20. Phase approval by (name) ________________________________

Many problems that crop up during or after construction or renovation usually stem from the fact that the consultant misinterpreted the owner's intentions. The consultant is not a mind reader, and needs to be told what is wanted or needed so that the consultant can tell the owner what can be done and how it can be accomplished.

The consultant, armed with this information, will develop a document that reflects the owner's requirements. Additionally, the consultant will review the estimated budget and anticipated completion schedule. He will determine any obvious incompatibilities, discuss these with the owner, and attempt to arrive at a mutual agreement.

In a full-service firm with both architects and engineers, the document completed will address architectural features and engineering concepts for all disciplines involved, for example: architectural, structural, HVAC, mechanical, electrical, plumbing, fire safety, and instrumentation/automation. The front matter of all disciplines will list all standard and applicable codes plus any other codes, guidelines, and regulations that impact the project.

The balance of the document consists of what is sometimes called an "outline specification." Some firms develop it into a more specific format and address design criteria for the project. For example, under HVAC, the following topics are covered:

1. Standards and applicable codes
2. Guidelines and regulations
3. Major code impacts
4. Standard operating schedule
5. Basic systems description
6. Systems evaluation
7. Design criteria
8. Interior load criteria
9. Expandability provisions
10. Flexibility criteria
11. Fuels availability
12. Energy inflation rate

This document addresses program requirements and design criteria. It restates the project scope; defines whether building and systems expansion provisions are considered; outlines outside environment as a future basis for inside HVAC design; establishes HVAC design levels for all spaces; defines the amount and type of lighting that will be the basis for design; notes emergency power requirement; UPS systems; lists fire suppression systems and fire alarm systems to be studied. This document, to be effective, must cover external items that impact on the project, address site and architectural features, list structural systems to be considered, note inside environment to be used for design, and so on, noting all features, conditions, and systems that fully describe the criteria to be used for the design of the project. The more complete the document, the better understanding both parties will have, and the less opportunity there will be for misunderstanding as the project progresses.

This document will reflect back the consultant's interpretation of the owner's intent. At the completion of this phase, the owner receives copies, must review them completely, and bring to the consultant's attention all discrepencies. When both parties agree on the contents of this document, the owner signifies approval in writing to the consultant and directs the consultant to begin the next phase.

Schematic Design Phase. This phase is handled differently by different firms. Frequently, where the predesign phase is not used, this becomes the first document developed and submitted to the owner for review and approval. In either case, the basis for design is addressed and developed in this phase. At the completion of schematics, some firms develop a minimum number of drawings that depict architectural concepts, while others develop $8\frac{1}{2} \times 11$-inch drawings that are bound into the document. These include such sketches as site plan, building concept, structural systems, mechanical systems, and distribution concepts, single-line electrical distribution, plumbing riser diagrams, and fire protection distribution. Whatever is necessary to fully illustrate the design concept should be included, but this will vary with the philosophy of consultants. This document, properly executed, forms the basis for design, and it will be used by the members of the project team for the remainder of the design effort.

The most important aspect of this phase is to establish the basis for design. For example, several structural systems are studied, appropriate sketches are developed to indicate the various trade-offs involved, and a cost per square foot is estimated. At the conclusion of this section a recommended system is identified. Since structure impacts on architecture, the recommendation is based upon total involvement. While one structural system may be more economical, its impact on other building systems may eliminate it from economic consideration.

This phase, for some firms, is a true economic evaluation of the building and systems. In order to accomplish the study effectively, building orientation consideration is essential, type and amount of exterior glass is evaluated, coefficient of heat transmission of walls and roof is established, and the type of fuels are included in systems economical evaluation. As you can see, this phase is probably the most important one of all as it establishes the basis for design. The next two phases merely document that which has been resolved, in such a manner that a contractor can construct the new building or accomplish the renovation. In some cases, this phase includes value engineering and life-cycle costing.

As was stated earlier, the schematics phase varies somewhat depending upon the established practice of the consultant. One firm develops what is termed as fundamental design standards. This document, in addition to establishing basis for design, contains the following:

1. Outline specification details for the work
2. Equipment selection list showing basic selection and approved equal
3. Typical details for installation of equipment

This additional information can be very beneficial to both the owner and consultant. For example, the owner can review the list of equipment proposed for the project and can delete, add to, or modify this list to best suit his needs, long before any significant effort is expended during the design documentation phase. This mutually agreed upon list of equipment will truly reflect the recommendation of the consultant and the needs of the owner. The design and final specifications will be accomplished around this list thereby eliminating surprises at the conclusion of the project. Since many owners prefer certain types of equipment for any number of reasons, it is well to recognize this need at an early stage of design and to appropriately address it.

Similarly, the inclusion of typical installation details shows the owner how the consultant plans to have this installation accomplished. The owner has an opportunity to review these details long before valuable time and money are spent by the consultant on final drawings. Again, depending upon the knowledge of the owner, he may prefer, based upon past experience, specific methods of installation.

All firms review the owner's anticipated project schedule and submit, as part of the schematic phase, a design and construction schedule indicating all essential milestones. Some firms provide a preliminary schedule during the program planning stage performed before the predesign phase. In this case this schedule is reviewed and updated as necessary. Also, at the conclusion of schematics, the consultant provides a construction cost estimate. The architectural and structural elements are usually detailed to an extent that permits a reasonable degree of quantity take-off. The mechanical and electrical systems are defined primarily by systems diagrams and the estimates are based on a unit cost approach. The most recent industry reports of productivity, union working rules, and availability of labor in the geographic location of the project are reflected in the pricing. The estimate contains probable escalation based on the scheduled start and duration of construction. This estimate also contains a contingency factor for auxiliary items not separately identified, plus a bidding contingency in the amount of approximately 15 percent with 10 percent for auxiliary items and 5 percent for bidding.

The consultant's objective is the development of construction documents; drawings and specifications, and, ultimately, the construction or renovation of a building, meeting the owner's requirements, complying with applicable regulations, guidelines, and codes, and capable of execution at a reasonable cost. This objective cannot be achieved unless the consultant has a clear and complete understanding of all the requirements and the constraints affecting the design. Much of this information is obtained from the owner, the rest from regulatory agencies, and some from investigation of site, utilities, drainage, climate, etc.

In order to accomplish the objective effectively, the following checklist identifies information required at the start of this phase and items to be reviewed and/or approved by the owner. In the space at the right indicate who is to provide or accomplish (O = Owner, C = Consultant.)

OWNER'S CHECKLIST—SCHEMATIC DESIGN PHASE

1. Approval of predesign phase ________
2. Land survey ________
3. Area map ________
4. USGS map ________
5. Topographical survey map showing
   a. Reference datum ________
   b. Water ________
   c. Sewer: sanitary ________
   d. Sewer: storm ________
   e. Culverts ________
   f. Flood plane ________
   g. Electric ________
   h. Gas ________
6. Aerial photo ________
7. Weather data ________
8. Soils boring report ________
9. Soils testing ________
10. Perculation test ________
11. Zoning ordinance ________
12. Agency review ________
13. Building code ________
14. Environmental regulations ________
15. Planning board submission ________

16. Permits data requirement
    a. Local ____________
    b. State ____________
    c. Occupancy ____________
17. Utilities interface ____________
    a. Electric ____________
    b. Gas ____________
    c. Water ____________
    d. Sewer ____________

The following covers items included in the schematics phase and requires review by the owner.

18. Architectural features ____________
19. Structural systems ____________
20. HVAC systems selection
    a. Chillers ____________
    b. Air handlers ____________
    c. Air distribution drawings ____________
    d. Terminal devices ____________
    e. Comfort design level ____________
    f. Fuels evaluation ____________
    g. Energy audits ____________
    h. Comparative cost summary ____________
21. Electrical systems selection
    a. Primary service voltage ____________
    b. Estimated building load, kva ____________
    c. Building distribution ____________
    d. Interior lighting levels ____________
    e. Exterior lighting ____________
    f. Telephone service ____________
    g. Emergency power ____________
    h. UPS systems ____________
22. Plumbing systems selection
    a. Domestic water demand ____________
    b. Domestic heating water demand ____________
    c. Domestic water heating costs ____________
    d. Chilled drinking water ____________
    e. Fixture selection ____________
    f. Food service connection ____________
    g. Riser diagrams ____________
23. Fire safety
    a. Building type of construction ____________
    b. Fire alarm system ____________
    c. Fire protection sprinkler, wet ____________
    d. Fire protection sprinkler, dry ____________

e. Halon fire suppression ________
f. Fire extinguishers ________
g. Hose cabinets ________
h. Booster fire pump ________
i. Smoke detection ________

24. Instrumentation/Automation
    a. Fire safety ________
    b. HVAC Automation ________
    c. Security ________
    d. Single line equipment diagram ________
25. Review schematic document ________
26. Delineate suggestions ________
27. Outline requests for change ________
28. Provide written approval ________

Preliminary Design Phase. Now that the schematic phase has been completed and the appropriate basis for design has been documented by the consultant and approved by the owner, it is time to move into the following phase. While this is frequently referred to as preliminary design, it is more accurately design development. This is where the consultant transfers to drawings the concepts developed to this point.

Construction drawings are usually drawn to an $\frac{1}{8}$ inch scale with $\frac{1}{4}$-inch scale or larger blow-ups of areas requiring greater detail. Preliminary drawings are drawn either at $\frac{1}{16}$-inch or $\frac{1}{8}$-inch scale, depending upon the size and complexity of the project, and to some extent as to the consultant's preference. The scale of the preliminary drawings warrant additional discussion.

Preliminary drawings at $\frac{1}{16}$-inch scale are essentially discarded after that phase is completed. They are used only for reference and a completely new set of drawings are developed for construction documentation. Preliminary drawings at the $\frac{1}{16}$-inch scale are a simple graphical presentation of the building configuration, equipment location, and systems distribution methods. Drawings are usually single-line floor plans indicating physical relationships. These drawings show space utilization and circulation relationship depicting the consultant's interpretation of the owner's requirements. This phase, when viewed as design development, is exactly that—the design is being developed.

During this phase there must be continuous interaction between consultant and owner. The owner must meet on a regular basis with the consultant and review the development of the drawings frequently to insure that his objective is being appropriately addressed. When these drawings are completed to the extent that all essentials are indicated and the owner approves them, these drawings are set aside for reference. A completely new set of drawings will be developed at $\frac{1}{8}$-inch scale that will be used for construction purposes.

Since this is the case in many instances, why not make preliminary drawings at the $\frac{1}{8}$-inch scale and reuse these for construction drawings? Probably the most important reason is because the building depicted on the construction drawings is frequently different from the one on the preliminaries. In this instance, the preliminaries could not be economically retained for use during the final design phase. Secondly, since the concept and configuration is often significantly changed during design development, it is impractical to keep revising these drawings when $\frac{1}{8}$-inch scale detailed drawings are used.

On the other hand, on smaller or less complex projects where there is early agreement on the building concept and space utilization, it is generally far more economical and design time can be shortened by doing preliminaries at $\frac{1}{8}$-inch scale. Where this method is used, the usual procedure is to conclude the preliminary phase as soon as the drawings are completed sufficiently to adequately describe the building, space utilization, circulation pattern, equipment location, and distribution methods. Then prints are made for owner review and approval. Once this is accomplished, the final phase of design is started utilizing the same set of reproducible drawings. All that is now required is to complete the documents by providing appropriate details, drawing necessary sections and elevations and developing enlarged scale blow-ups to define critical areas. From this it is readily apparent that significant cost and time savings could be realized. The method selected for preliminary documentation depends upon the size and complexity of the project, to some extent upon the capability of the consultant, and the owner's requirements.

Some owners, particularly those with adequate facilities departments, establish their own requirements for preliminary documentation. The following is developed to show typical requirements for preliminary documentation:

PRELIMINARY DOCUMENTATION REQUIREMENTS

Prepare preliminary drawings and specifications depicting the project in adequate detail to reflect owner's requirements and sufficiently detailed for quantity takeoff estimating: (Drawing size and title block is determined by owner)

1. Architectural
   a. Building location
   b. Site access, roads, parking areas, grading, and secondary structures
   c. Utility rights-of-way
   d. Surface drainage systems
   e. Floor plans showing primary partitioning
   f. Exterior elevations
   g. Building sections
2. Civil
   a. Site layout
   b. Sanitary sewers
   c. Surface drainage and storm sewers
3. Structural
   a. Substructure showing footings, foundations, piers, pilings
   b. Superstructure showing beams, girders, columns, and floor thickness
   c. Structural walls and supports
   d. Roof thickness
4. Mechanical
   a. Location of shaft spaces and mechanical rooms
   b. Size and location of major equipment

   c. HVAC distribution, including pipe and duct sizing—single line
   d. Terminal equipment
5. Electrical
   a. Location of electrical rooms and spaces
   b. Size and location of major equipment
   c. Emergency and exit lighting
   d. Emergency generator
   e. General lighting
   f. Local distribution panels
   g. Site utilities
   h. Site lighting
   i. Communications system
   j. Telephone service, rooms, and closets
6. Plumbing
   a. Size and location of major equipment
   b. Location of fixtures
   c. Roof and floor drains
   d. Riser diagram for water, sanitary, and storm systems
   e. Single line plans for water distribution and storm and sanitary systems
7. Fire protection
   a. Size and location of fire pump
   b. Risers and stand pipes
   c. Sprinkler distribution system
   d. Hose cabinets
   e. Halon system
   f. Area to be sprinklered
8. Instrumentation/Automation
   a. HVAC automation single line diagram
   b. Security single line diagram
   c. Fire alarm single line diagram
9. Project milestone schedule depicting the balance of design and construction effort
10. Preliminary (outline) specifications
11. Preliminary construction cost estimate for separate line items for all major pieces showing quantities and unit prices of equipment, systems, installations complete with appropriate backup and supported by a brief description of the estimating assumptions or sources of information as deemed necessary.

Since the owner has been meeting on a regular basis with the consultant and reviewing the design and documentation during this entire phase, the completed preliminaries should correctly reflect the owner's intent. This being the case, the extent of final review could be less involved and owner approval could be provided more rapidly. In fact, in some cases, the consultant is directed by the owner to continue into the construction documentation phase. In either instance, the consultant requires written approval to proceed.

Construction Contract Documentation Phase. From an owner's point of view, this phase may be the simplest of all, while for the consultant it is the most crucial. The owner may desire to be heavily involved or merely monitor progress at his discretion. In any case, the consultant will hold regularly scheduled discipline coordination meetings, conduct continuous estimating to assure component budget adherence, and monitor design progress to insure completion on schedule.

One of the major errors committed by nonprofessional owners is that they expect unrealistic levels of performance in the preparation of contract documents. It would be reasonable to state that *complete drawings and specifications* do not exist. According to dictionary definitions the word "complete" means not lacking or incomplete in any respect, i.e., perfect. Drawings and specifications are not an end in themselves but merely the documents by which the consultant conveys the design requirements to the contractor so that the project can be constructed. In other words, don't expect the impossible and you won't be disappointed.

At about 50 percent of completion of the construction contract documentation phase, the consultant will:

1. Submit for owner's review copies of general conditions and supplementary conditions of the specifications;
2. Review with owner type of construction contract contemplated;
3. Develop, with owner's input, a list of potential contractors;
4. Determine what items are to be furnished by the owner;
5. Obtain owner's requirements for phased occupancy.

The consultant will issue drawings to the owner for review and comment, if necessary, at regular intervals during this phase. Usually, at 90 percent of completion of the drawings, the owner will receive a complete set for final review, a final construction cost estimate, and a final set of specifications for review.

At this stage of completion, the consultant will:

1. Obtain owner's instructions on insurance and bonds;
2. Review list of potential contractors;
3. Assist owner in filing documents for approval and permits;
4. Establish, with owner's input, a list of testing agencies necessary for the project;
5. Assist owner with filing documents required for approval of authorities having jurisdiction over project;
6. Establish final details for date and location for contractors to pick up drawings and specifications for bidding purposes and date of bid receipt.

Bidding or Negotiating Phase. The degree of involvement by the consultant could vary significantly depending upon the agreement between the owner and consultant. Therefore, what follows is the general rule but responsibilities may shift from owner to consultant as mutually agreed upon:

Owner's Responsibility

1. Advertise for bids
2. Distribute bid documents to bidders

3. Attend prebid meeting
4. Review addenda
5. Receive bids
6. Open bids
7. Select contractor
8. Issue letter of intent to selected contractor
9. Prepare and execute construction contract

Consultant's Responsibility

1. Schedule prebid meeting with bidders a minimum of two weeks before scheduled receipt of bids
2. Attend prebid meeting
3. Prepare and distribute addenda
4. Attend bid opening
5. Tabulate and analyze bids
6. Assist owner in the selection of a contractor
7. Assist owner in preparing the construction contract.

As you can see from the above, the consultant's role in this phase is generally one of providing assistance to the owner.

Construction Administration Phase. This is the phase where many nonprofessional owners encounter the most difficulty because they expect the consultant to direct and supervise the contractor. This is normally not the case. The owner has two separate contracts; one with the contractor and the other with the consultant.

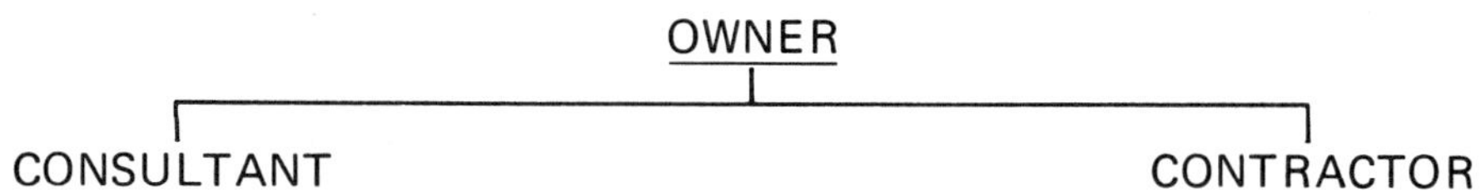

Generally, unless otherwise provided in the agreement with the owner, the consultant will serve as a representative of the owner and consult with and advise the owner. The degree of participation and extent of responsibility depends upon the size and complexity of the project.

Owner's Checklist—Construction Administration Phase

1. Attend preconstruction meeting
2. Obtain bonds and insurance certificates from contractor
3. Review and approve subcontractors
4. Review construction schedule submitted by contractor
5. Attend all job meetings

6. Review and approve bulletins, cost proposals, and change orders
7. Issue change orders
8. Process regular payments to contractor
9. Apply for permanent utility services
10. Arrange for authorities having jurisdiction to inspect site as required
11. Obtain certificate of occupancy
12. Make final inspection of project

The consultant, as the owner's representative, will assist the owner in the above and will attend all job meetings; issue minutes of meetings; make periodic visits to the site; review shop drawings and other submissions; prepare and issue bulletins; review cost proposals and prepare change orders; issue progress reports; prepare punch-lists; review as-built drawings and maintenance manuals; obtain warrantees and guarantees.

This concludes the normal work provided by consultants under basic services and is usually included in the basic fee schedule.

Additional Services

The range of services that are additionally available is almost unlimited and depends entirely upon the needs of the owner. The consultant selected will provide details of the services that he furnishes or obtains from others. These services are usually negotiated separately and paid for by the owner as a multiple of the consultant's Direct Personnel Expense.

The following is a partial list of additional services:

1. Preparation of applications and supporting documents for government grants, loans and advances.
2. Making drawings from field measurements of existing construction when required for planning additions and alterations.
3. Revising previously approved drawings or specifications to accommodate owner's changes.
4. Preparation of detailed renderings, exhibits and/or scale models for the project.
5. Providing prolonged construction administration should the construction contract time be exceeded by more than 25 percent due to no fault of the consultant.

SELECTING DESIGN CONSULTANTS

The first step toward a successful conclusion of a given project is the selection of an appropriate consultant whether an architect, engineer, construction manager, or design/builder. This is a guideline for the selection of a consultant and, similarly, consultants have their own guidelines for the selection of a client (owner). While what follows addresses your concern only, be assured the consultant will be as careful as you are in the selection process. When you have selected the consultant best suited to the particular project needs, you will establish a relationship mutually beneficial to both parties.

Owners have been known to use a variety of methods for the selection process that range all the way from the company's executives' personal preference to the highly sophisticated approach where several prequalified consultants are asked to make full-blown presentations outlining their experiences on similar projects, explaining their services, addressing the method of handling the project, and introducing the key members of the team to be assigned.

Since it is unfair and expensive to consultants to expect full-blown proposals without first prequalifying and limiting the number to just a few, you should prequalify consultants so that you can obtain an optimum match to your project. Unless you are intimately familiar with consultants, the best approach is through the use of standard analysis forms. The ones included here are the result of years of experience with consultants. Most consultants will gladly complete these or similar forms and may even supplement them with their own expanded format that includes such additional information as:

1. List of similar projects by name, construction cost, and year of completion
2. List of major honors and awards received for design
3. List relationship between final construction cost estimate and actual construction cost
4. List previous experience of the individuals on the project team

This, coupled with information on the analysis forms, should put you in a position to make a direct selection for interview or, at least, narrow the list to a select few.

As was stated earlier, you must be able to define the project in order to inform the consultants of your intention. This is best accomplished by completing the owner's checklist—Predesign Phase. Attach this to the anaysis forms and you are ready for mailing to consultants. Even this mailing should be preselected to the extent of size and similar experience. One owner selects those consultants that are of a size where the project load does not exceed 50 percent of the consultant's capacity.

After the consultants have responded, you have sufficient information to proceed with the selection process. Select those few, usually not more than three, and set up meetings at your offices where the consultants will conduct their formal presentation. Since this proposal presentation is important to both you and the consultant, give them sufficient time to prepare for the presentation. In most cases, two weeks will be adequate. Also, establish a time limit for the presentation. Usually an hour is enough time but try not to exceed two hours. A short presentation is preferred as it is usually more to the point and permits you the best opportunity to evaluate the content.

There are several key checks you need to evaluate the consultant.

1. Check the responsibility of the main presenter. Is he a key member of the firm or just one of the middle managers? This establishes the level of importance they attach to the presentation.
2. Does the proposal address your particular project or is it of a general nature?
3. Determine how well the presenters have responded to your needs.
4. Check the size and types of projects the consultant discusses in establishing his past experience. Is it relevant to your project?

CONSULTING SERVICES ANALYSIS FORM NO.1

Firm Name: ______________________________

Address: ______________________________

Name of two principles

	Name	Title	Education by degrees	Year graduated
1.	____	____	____	____
2.	____	____	____	____

Number of personnel by discipline

____ Administrative	____ Draftsmen
____ Architects	____ Designers
____ Landscape Architects	____ Economists
____ Civil Engineers	____ Estimators
____ Electrical Engineers	____ Interior Designers
____ Instrument Engineers	____ Planners
____ Mechanical Engineers	____ Specification Writers
____ Plumbing Engineers	____ ____________
____ Soils Engineers	____ ____________
____ Structural Engineers	____ ____________
____ Surveyors	____ ____________
____ Total Professionals	____ Total Personnel

Range of Professional Service Fees (Past Year)

____ Less than $100,000	____ $1.0 million to $5.0 million
____ $100,000 to $500,000	____ $5.0 million to $10 million
____ $500,000 to $1,000,000	____ More than $10 million

Summary of Professional Services Fees (Past 4 Years)

19___ ________ 19___ ________ 19___ ________ 19___ ________

Figure 5-1. Consulting Services Analysis Form No. 1.

5. A well-organized presentation indicates a well-disciplined consultant.
6. Disregard all mention of fees and fee schedule unless this information is requested by you. Never select a consultant purely on the basis of fee alone.
7. Finally, pay careful attention to the presentation and permit the consultant to proceed as planned. Interruptions are distracting and tend to disrupt the flow of the presentation. Save your questions for the informal part of the session.

FEE SCHEDULES

As we noted above, never select a consultant on the basis of fee as the sole criterion. It does not follow necessarily that the higher the fee the better the quality of the consul-

CONSULTING SERVICES ANALYSIS FORM NO. 2

Profile of Firm's Project Experience (Past 5 Years)

	Profile Code	Number of Projects	Total Project Cost, in Thousands
1.	________	________	________
2.	________	________	________
3.	________	________	________
4.	________	________	________
5.	________	________	________

Experience Profile Code

1. Airports & Terminals
2. Commercial Buildings
3. Chemical Processing
4. Educational Facilities
5. Healthcare Facilities
6. Industrial Buildings
7. Office Buildings
8. ________
9. ________
10. ________

Profile of Firm's Services (Past 5 Years)

	Profile Code	Number of Projects	Total Project Fees, in Thousands
1.	________	________	________
2.	________	________	________
3.	________	________	________
4.	________	________	________
5.	________	________	________

Services Profile Code

1. Site Selection
2. Project Evaluation
3. Master Planning
4. Building Programming
5. Design Services
6. Construction Administration
7. Construction Management
8. Assist Move & Occupancy
9. Full Services (All Above)
10. ________
11. ________
12. ________

Figure 5-2. Consulting Services Analysis Form No. 2.

tant; he may just have a higher overhead or less efficient performance. Similarly, a lower fee does not mean a better deal, nor does it indicate a more efficient organization. The fee is what you pay for the service, that and nothing more.

There are two methods generally used by most consultants to develop their fee. In one case, the fee is established as a percentage of the estimated construction or renovation cost. In the other, the fee is based on the actual anticipated hours of work involved. The percentage-of-construction cost fee is most frequently used on new building construction and large remodeling projects.

The rate of the fee is inversely proportional to the construction cost and directly proportional to the complexity of the project. Just remember that a consultant must turn a profit or will not be in business very long. Secondly, the consultant must remain competitive or will never obtain work. Finally, he must do quality work or he will never be given repeat work from the same owner.

On smaller construction projects or building renovations, many consultants determine the number of hours used to complete the construction contract documents. To this they add the anticipated hours required for site visits, construction administration, meetings, plus reimbursable expenses.

Some consultants use the cost-plus-fixed-fee method, particularly when the project is not clearly defined. This is particularly beneficial when you desire to contract for services on a piecemeal basis, i.e., you would like to obtain information and design on one phase at a time. This method offers excellent advantages when you need assistance early on in developing the scope of your project. Most consultants charge from $2\frac{1}{2}$ to 3 times payroll costs.

Some consultants submit cost proposals of 2 to $2\frac{1}{2}$ times direct personnel expenses (DPE). This includes actual salaries plus costs for Social Security (consultant's share), employee's liability, comprehensive general liability (bodily injury), property legal liability insurance, and payroll benefits.

There are also expenses reimbursable at actual costs. These include: transportation, travel and living expenses; reproduction costs; premium portion of overtime pay incurred at owner's request; outside consulting services; license, permits, testing, and inspection fees incurred because of the work.

You may require additional assistance during the construction administration phase. This is *additional services* and is usually negotiated separately. This cost is above the fee for basic services. The range of costs for these services varies widely depending upon the degree of involvement anticipated. This phase is often performed on a cost-plus-fixed-fee basis.

SUMMARY

The process of selecting a consultant is rather simple and straightforward, but selecting the one best suited to your project is a little more difficult. The end result of a choice of consultant is a quality set of construction documents, and the better the drawings the more nearly perfect will be the building.

1. Define the project to the extent necessary so that the consultant will understand your intentions.
2. Preselect a number of consultants that appear to be appropriate in size of operation and experience.
3. Request these consultants to respond by completing the Consulting Services Analysis Forms No. 1 and No. 2.
4. Select several consultants, not more than three, to make presentations.
5. Their presentations should address your specific project.
6. Carefully evaluate each presentation and associated proposal.
7. Select the one best suited to your requirements.

8. Contract for the services needed and make sure that the contract is well defined.
9. After you have made your choice and signed a contract, cooperate and work with the consultant as the quality of the final construction drawings and the successful completion of the project depends as much on your cooperation as on the consultant's skill.
10. Never select a consultant solely on the basis of fee alone. The fee is what you pay for the service, that and nothing more.

6. Cost Estimating Techniques

The ability to estimate construction/installation and renovation projects is a valuable asset in building and facilities management. In addition to knowing how to estimate, it is important to know the degree of accuracy required at the various stages of the project. Further, estimates cannot be made more accurate than the degree of definition of the project. For example, at the conceptual stage, a square-foot cost will usually provide sufficient accuracy to determine the feasibility of the project. Whereas, after the contract drawings and specifications have been completed, the contractor will be able to determine the cost of the project to such accuracy that he would be willing to agree (contractually) to complete the construction and installation at a fixed price.

There are several levels of estimating accuracy for any project depending on the definition of the project. The contractor's estimate—that's all his cost proposal is—must be very accurate because he gambles the profitability of his company on that estimate.

Facilities managers are frequently required to provide conceptual estimates at an early stage of development of a project. Depending upon the project, he may be called upon to review cost estimates submitted by design consultants, the final cost proposal submitted by contractors, and various cost estimates anywhere in between. In cases where design consultants are involved, they provide cost estimates at the end of various phases consistent with the definition of the project at that stage. While they make very effort to develop an accurate estimate, the numbers they provide are just that, an estimate. In other cases, particularly smaller construction and renovation projects, facilities managers may handle all aspects of estimating and reviewing contractor's cost proposals.

Some people have a much better aptitude for the mechanics of estimating than others, but almost anyone can develop such skills to the point where they will feel comfortable in any situation. Once skill in this area has been properly developed, it is necessary to continually update the variables; rising labor and material costs.

TYPES OF ESTIMATING

There are generally three levels or degrees of accuracy of estimating that cover most all situations. The first (with the least accuracy) is the conceptual, schematic, or square-foot cost estimate. Design consultants usually refer to this as a conceptual cost estimate while the nonprofessionals call it a ball-park guestimate. The next level is the unit-cost estimate, referred to by design consultants as the preliminary estimate, as this is provided by them at the conclusion of the preliminary drawings and specifications. The final estimate provides the most accurate form of estimating because it requires a complete material take-off from complete architectural and engineering drawings and specifications.

Conceptual Cost Estimates

This is actually an educated estimate based on actual costs of similar projects of similar type of construction. Generally, one does not visualize a building of a certain size and arbitrarily assign a cost to the whole building, although a thoroughly experienced professional can establish a fairly accurate estimate even in this manner. One must visualize a particular type of building of a certain size in a specific area and then determine the estimate based on past construction cost on a dollars per square foot basis. The *similar* type of construction is essential because comparing a poured concrete building with one of precast concrete plank on block bearing walls, or with a steel-frame curtain panel wall building, can lead to serious error. Further, a sophisticated central chiller plant HVAC system will certainly cost more than a packaged rooftop HVAC system. Also, an office building will cost significantly less than a hospital. Therefore, it is necessary in this case to compare similar types of buildings with like construction and equivalent HVAC systems.

In any type of estimating, the greater the number of individual line items compared, the greater the degree of accuracy. From this it is obvious that the first step must be to identify as many separate line items as possible consistent with the degree of development of the concept of the project. Then establish, by whatever means appropriate to the project, a cost per square foot for that line item. Next, multiply this line item by the number of square feet associated with the item. Finally, add up all the line items to obtain the total cost of the project. For example:

Line Item 1 × \$/sq ft × sq ft = \$ __________
Line Item 15 × \$/sq ft × sq ft = \$ __________
Total \$ __________

This sounds simple for new buildings, but what about renovating/remodeling existing buildings? That is covered in the next chapter.

There are several techniques that can be used to develop a list of individual line items. Also, the degree of accuracy desired at this stage of development will determine the number of line items required. For example, the cost of a new building can be roughly estimated based simply on the cost per square feet for the entire building. This is usually satisfactory for talking purposes only. A much more reliable method will identify the major items of construction. First, the basic components of the structure are identified then these are subdivided as required. Next, the various construction trades are listed. Finally, all special requirements are identified. It is important to note that the area of square feet associated with a particular line item must relate to that line item. For instance, when a building is partially sprinklered, only that area is used to determine the cost of the sprinkler system. This concept applies to air conditioning, lighting, etc.

Since site work costs are unique to the project, it is usually identified differently and estimated accordingly. In cases where the site has not been selected, the site work estimate is excluded. For purposes of this discussion, site work is included in the checklist. It is intended as a guide to illustrate the items usually considered.

Estimators usually use back-up worksheets that are attached to the summary to show how costs were developed. For example, in order to determine the cost of site paving, an estimator will have to know the amount of cars that will be parked at the site. He may use 300 square feet per car plus 15 percent for normal roadways. Access roads will be considered separately. His worksheet may resemble the following:

100 cars × 300 sq ft/car =	30,000 sq ft
15% for roadways	4,500 sq ft
Access road, 24 ft × 500 ft =	12,000 sq ft
Total	46,500 sq ft
@ $2.00 sq ft material & installation	$93,000
2,000 ln ft asphalt curbing @ $2.00	4,000
Total	$97,000

This total cost is then transferred to the appropriate line item and the worksheet is numbered, identified, and set aside for attachment to the cost estimating checklist.

CONCEPTUAL COST ESTIMATING CHECKLIST

Line Item	*Cost Estimate*
Site Work	
Demolition	________
Grading	________
Drainage	________
Paving	________
Landscaping	________
Fencing	________
Water service	________
Sanitary sewer	________
Storm drainage	________
Fire protection	________
Electrical substation	________
Electrical distribution	________
Site lighting	________
Building	
Foundation	________
Structural system	________
Exterior walls	________
Glass & glazing	________

Line Item	Cost Estimate
Roofing	________
Interior walls	________
Partitions	________
Painting	________
Floor covering	________
Ceiling	________
Mechanical	
Plumbing	________
HVAC	________
Sprinkler system	________
Automation	________
Electrical	
Substation	________
Distribution	________
Lighting	________
Specialties	
Elevators	________
Kitchen equipment	________
Fire pump	________
Emergency generator	________
Fire alarm system	________
Telephone system	________
Communications (paging)	________

Many components of a building can be estimated on a square-foot basis. Some more experienced estimators start with a square foot cost for the building and then determine electrical, mechanical, and structural costs on a percentage of the total cost. Since the percentage-of-total costs vary with the particular building, it is important to know the type of building under consideration. For example, plumbing percentage for a hospital could be as high as 10 percent of the total cost while for an office building it frequently is under five percent. This variation will also apply to electrical and mechanical systems. The structural cost will not vary as much with the type of occupancy as with the type of structure.

For the various components of a building that can be estimated on a square-foot basis, the following will be helpful:

Square Foot Cost Estimating Checklist

Line Item	Area, Sq ft	×	\$/Sq ft	=	Estimate
Structural system	________		________		________
Exterior walls	________		________		________
Glass & glazing	________		________		________
Roofing	________		________		________
Interior walls	________		________		________

(*continued*)

Square Foot Cost Estimating Checklist

Line Item	*Area, Sq ft* ×	*$/Sq ft* =	*Estimate*
Partitions	________	________	________
Painting	________	________	________
Floor covering	________	________	________
Ceilings	________	________	________
Electrical	________	________	________
Plumbing	________	________	________
Sprinkler system	________	________	________
HVAC	________	________	________
Automation	________	________	________

For the inexperienced estimator, some of the above items will create problems; for example, the structural system. In the schematic design phase (conceptual design) the consultant has developed a building concept. In fact, he has established the size of the building in square feet of area and the number of levels (floors) necessary to satisfy the project scope. In order for him to determine the most economical structural system that best suits the design criteria, the consultant may evaluate several bay sizes and will evaluate structural systems.

The following information is taken from an actual project. It shows how the consultant arrived at the square-foot cost of the structural system and you could use the same technique. The bay size that best suited the owner's requirements was determined to be 24 × 40 feet. The building was to be 100,000 square feet total with four floors. Three structural systems were compared: steel frame; reinforced concrete with one-way joist; and precast concrete double-tees. While the precast was more economical, the reinforced concrete required 11 inches less depth of construction. The 44-inch reduction in building height more than offset the cost differential and the reinforced concrete system was selected.

A typical bay was drawn for each system for the evaluation. Items common to all schemes were excluded so that the structural systems could be properly compared. This typical bay was then reduced to quantities on a square foot basis and was tabulated at shown below:

Cost of Structural System

ITEM	QUANTITY PER SQ FT	UNIT COST	TOTAL
Joists and Beams	0.03 cu yd	$300.00	$9.00
Columns	0.005 cu yd	350.00	1.75
Foundations & Caisons	0.004 cu yd	325.00	1.30
Rebars	7.6 lb	0.25	1.90
Total Cost per sq ft			$13.95
Plus 8.8% Escalation for Construction Period			1.06
Total			$15.01

The contractor's bid for this work was $15.38, showing an accuracy of estimating for this scheme to be around 2 percent.

Exterior walls, glass and glazing, interior walls, and partitions must be broken down to some convenient size to determine the cost-per square foot, but these must now be multiplied by the appropriate area, for instance, multiply the exterior wall cost by the square foot of exterior walls and not the square foot of the building.

Painting relates to the surface area to be painted and not the area of the building. When both sides of a wall or partition are to be painted, the area of both sides must be used.

In some cases, it may be appropriate to address lighting costs as an individual line item, whereas in others, it could be excluded and grouped in with the total costs for electrical work. The cost of such items as roofing, ceilings, floor coverings, and brick and block work, glass and glazing can be obtained from local contractors to provide the degree of accuracy required for this phase of estimating.

Items like electrical, plumbing, sprinkler systems, HVAC, and automation are usually most difficult to handle for the inexperienced estimator. Consultants develop histories based on past projects and are able to quickly and accurately estimate costs. In fact, many consultants can establish conceptual cost estimates in this manner within about 10 percent. The larger the reference history, the more accurate the estimate.

Whenever you are in doubt about the costs used for the above, try to obtain the necessary information from local contractors. Most contractors will gladly help you as they feel this creates good will.

It is also valuable to know the percentage of total costs. As stated before, some of the items vary with the type of construction, while others vary with type of occupancy.

For your information, the following was taken from the project mentioned above. It is a 100,000 square-foot, four-floor, fully sprinklered building.

Percentage-Of-Total Cost

LINE ITEM	% COST
Plumbing	2.9
Sprinkler system	2.7
HVAC	17.8
Automation	2.75
Electrical	8.3
Structural	15.5
Roofing	1.6
Acoustical ceiling	1.5

As was stated earlier, these percentage costs could vary significantly with the type of occupancy, for example:

LINE ITEM	HOSPITAL	OFFICE
Plumbing	7.0 to 11.0%	5.0 to 9.0%
Electrical	10.5 to 15.0%	8.0 to 12.0%
HVAC	12.0 to 20.0%	9.0 to 15.0%

In summary, a conceptual cost estimate is based on the building concept or, in the vernacular of the consultant, it is a schematic estimate. It is useful in determining the budget estimate.

Preliminary Cost Estimate

This estimate is developed by consultants at the completion of preliminary drawings and specifications. It is also known as a unit-cost estimate. The consultants develop this estimate using their preliminary drawings, usually made at $\frac{1}{16}$-inch scale, depicting the building, systems, and equipment. Sometimes $\frac{1}{4}$-inch blow-ups are developed to define in greater detail such areas as mechanical rooms, electrical rooms, building structural systems, and appropriate sections and elevations are shown as required.

In cases where you make your own drawings, you should be able to develop this level of estimate. This is a refinement of the estimate previously outlined and provides a greater degree of accuracy. At this stage, systems and equipment are defined in greater detail and their individal costs must be included. Most importantly, of course, is to identify and list every system and all equipment. The more line items included, the greater the degree of accuracy of the estimate. The major fear of all estimators is that something is overlooked or missed.

For drawings to be useful for unit-cost estimating, they should show all equipment, pipes, ducts, lighting fixtures, sprinkler heads (or cross-hatching to identify sprinklered areas), walls, partitions, etc. From this it is possible, for example, to determine the lineal footage of pipes, and then this is multiplied by the cost per lineal foot. As you can see, this method can provide a more accurate estimate than the cost-per-square-foot or percentage-of-total-cost methods.

Since construction work is performed by various trades, the estimate should similarly be developed and then there will be an opportunity for comparison. This separation and identification of the different sections of estimates should be carried through for all three levels of estimating.

The plumbing estimate is no longer a single line item, as it now includes all fixtures, equipment, and material. The estimating checklist is expanded and becomes more comprehensive. The list should be as complete as possible consistent with the level of development of the drawings and specifications. Equipment and fixtures are usually noted by quantity and multiplied by unit cost, pipe is noted by lineal footage, sheet metal duct by pounds, concrete by cubic yards, parking surface by square yards, ceilings by square feet, etc.

One common error committed by estimators is to forget contractor markup. For example, a sheet metal contractor may tell you that he charges $2.00 per pound of duct installed. Remember that since he is a subcontractor to a mechanical contractor who is a sub to a general contractor, these two contractors' markup must be included to reflect the true cost to you.

This method and format is best illustrated by the following example: (Unit prices are for illustrative purposes only and do not reflect actual costs)

Sample Unit-Cost Estimating Checklist

LINE ITEM	UNITS	QUANTITY	UNIT COST	TOTAL COST
Water Closets/ Carrier and Rough-in	EA	10	$ 900.00	$9000
4″ Cast Iron Pipe	LF	100	8.00	800
Ceiling, 2 × 4 Fiber Board Tile	SF	1000	1.10	1100
Concrete, Slab-On-Grade/Mesh	CY	20	175.00	3500
Bituminous Paving W/Crushed Stone	SY	1000	7.50	7500
Sheet Metal Duct	LB	1000	2.50	2500
Insulation, 1″ Fiber Glass	SF	200	2.75	550
Roof Top AC, 10 Ton Single Zone, Elec, Heat	EA	1	6000.00	6000

Abbreviations

EA = Each	CY = Cubic Yard
LF = Lineal Foot	LB = Pounds
SF = Square Foot	LY = Square Yard

Unit-cost estimating is a rather straightforward method for determining the cost of any project. Once you have identified all individual pieces of equipment, systems, and material in the project, and have determined the installed unit costs, all you have to do is extend the number and add up the costs. Remember to use appropriate markup for sub and general contractors. Then you add in a contingency factor for miscellaneous items not generally identifiable. Also include a percentage for labor and material escalation to project the estimate from the date of the estimate to at least the midpoint date of construction. A 10 percent contingency factor is usually sufficient, but the escalation percentage depends upon labor and material, and this could vary significantly depending upon the economy of the times. Finally, add about 5 percent for bond premiums, licenses, mobilization, and general conditions for the general contractor.

The first step in making a unit-cost estimate is to develop a checklist identifying as many separate line items as practical and necessary. While no list can include every item for every project, the following should serve as a guide. You may need to add to or delete from the following:

UNIT-COST-ESTIMATING CHECKLIST

Site Work

Site grading
Bulk excavation and backfill
Surface drainage
Subsurface drainage
Landscaping
Site Concrete
Parking area paving

Building

Foundation excavation and backfill
Footings and foundation walls
Concrete
Rebar and wire mesh
Concrete masonry walls
Brick masonry
Waterproofing
Miscellaneous iron
Roofing
Metal flashings
Metal doors and frames
Wood doors
Entrance doors
Glazing
Skylights
Access doors
Steel curtain walls
Lath and plaster
Dry-wall partitioning
Acoustical ceilings
Resilient tile flooring
Carpeting
Vinyl wall covering
Painting

Plumbing

Water closets
Urinals
Lavatories
Cast iron piping
Copper piping
Piping insulation
Electric water coolers
Hot water heaters
Roof drains
Floor drains
Drain piping

Electrical
- High voltage service
- Transformers
- Motor control centers
- Distribution panels
- Feeder conduit and wiring
- Safety switches
- Receptacles
- Telephone outlets
- Lighting fixtures
- Security and fire alarm systems
- Emergency generator
- Miscellaneous items

HVAC
- Air conditioning chillers and pumps
- Cooling tower and pumps
- Air handler units
- Perimeter radiation units
- Sheetmetal ducts and insulation
- Grilles, diffusers, and registers
- Chilled and hot water piping and insulation
- ATC systems
- Miscellaneous items

Fire Protection
- Fire pumps
- Fire detection and alarms
- Sprinkler heads, pipe, and fittings
- Fire extinguishers
- Water connection and site work
- Site fire mains and PIVs

Special Equipment and Construction
- Elevators
- Dock levelers
- Pneumatic conveyors
- Kitchen service equipment

Contingency
Escalation
General Conditions

This list is far more extensive than that required by a professional estimator. The additional items are included to assist you in evaluating your particular project.

The next important consideration is the determination of the various cost factors for the different line items. For example, you will need to know costs per square foot, cost per lineal foot, cost per cubic yard, etc. These factors can be obtained from experience, from contractors in the area or from estimating manuals. There are several excellent sources available to assist you, in fact, one such manual has been on the market for four decades. These manuals are a valuable source for any type of estimating at all. Most show complete identification of all items of construction and installation, and explain in detail how to use the manual. Since these manuals are issued annually, updating all relevant costs, no attempt will be made here to show these costs as they usually vary from year to year.

Final Cost Estimate

This is the most accurate estimate of the three discussed. It is performed by consultants at about 90 percent completion of construction contract drawings to indicate the actual cost of the project. In the event that this estimate indicates a cost beyond the owner's budget, changes in design can be accomplished to reduce the cost to within budget limits or, if the owner desires, the owner's budget can be increased to cover the level of design depicted in the contract documents.

This estimate is known as a bid estimate when performed by contractors as it forms the basis for their bid on the project. It is also known as a take-off estimate because all material and equipment is *taken off* the drawings and specifications. While the contractor's estimate may be no more accurate than the consultant's, the contractor will be bound to perform all work shown on the drawings and described in the specifications. The contractor's profit will ride on the firm's ability to accurately estimate the actual cost of the project. Keep in mind that contractors are not immune from error. This is important to remember when reviewing contractor's bid proposals. While it is the customary practice for an owner to accept the lowest bid, this may sometimes create serious problems. Contractors have been known to go into bankruptcy trying to complete a project for a cost where serious error or ommission was involved.

Take for example the following bid proposals submitted by five contractors for a hospital renovation project estimated by the consultant to be $407,000.

CONTRACTOR	BID PROPOSAL
A	$349,000.
B	401,275.
C	405,300.
D	412,999.
E	525,000.

Investigation revealed that Contractor E was in the midst of a large construction project. This was a courtesy bid because Contractor E did not want to be stricken from the prequalification list by refusing to offer the bid. Contractor E felt safe at that number in the event it became low bidder.

On the other hand, Contractor A, while experienced in new hospital work, had minimal experience with hospital renovation. Contractor A did not take into account the condition that all normal services at a hospital must not be disrupted during the period of the contract. These conditions included in the specifications stated:

- Corridors shall not be obstructed at any time.
- Material will be stored in the basement and brought to the construction area in such quantities as can be installed immediately.
- All construction areas will be broom clean after each working day.

The owner had three options:

1. Accept the proposal.
2. Reject the proposal.
3. Call in the contractor to review the proposal.

The owner elected to call in Contractor A before awarding the contract. In this case, the contractor stated that the special conditions had been overlooked and elected to withdraw the bid. The contract was then awarded to Contractor B.

The digression from estimating was to point out your usual involvement at this phase. Facilities managers and building owners and managers do not normally get specifically involved in take-off estimating, but you should understand what it is all about—who does what, including you.

Estimating from completed construction contract drawings and specifications is somewhat easier that from schematics and preliminaries for the experienced estimator and is more difficult for the others. It should be easier for everyone because all equipment is identified and sized, all material is shown, and the building is fully dimensioned. Appropriate sections are cut and elevations are shown, and critical areas are blown up to a larger scale for ease of identification of installation anticipated.

The experienced estimator leaves nothing to chance, he needs to know that he has fully covered the project. First, the estimator takes the specification format that his office uses and develops an estimating checklist based on that format. You should do the same in cases where you are called upon to provide this level of estimation. Generally, specifications are divided into 17 divisions, as follows:

1. General conditions
2. Site work
3. Concrete
4. Masonry
5. Metals
6. Wood work
7. Thermal and moisture protection
8. Windows and doors
9. Finishes
10. Specialties
11. Equipment
12. Furnishings
13. Special systems
14. Conveying systems
15. Mechanical

16. Electrical
17. Testing and balancing

This is further subdivided, for example, under Plumbing, as part of Division 15. Mechanical:

Plumbing—Inside Building (one concept is for plumbing to be responsible for all work within the building to a point 5 feet outside the building)

a. General provisions
b. Excavation and backfilling
c. Hot and cold water pipe and fittings
d. Hangers and accessories
e. Valves
f. Expansion joints
g. Isolation joints
h. Insulation
i. Pumps
j. Fixtures and trim
k. Water heaters
l. Water coolers
m. Soil, waste, and vent piping
n. Sump pumps
o. Sewage ejectors
p. Grease interceptors
q. Lift stations
r. Water treatment
s. Testing and adjustment

Again, it is not possible to list every item for every situation, for example, a hospital project will include such systems as: oxygen, nitrogen, nitrous oxide, compressed air, and vacuum systems. This list should be developed to suit your particular project. Section j, Fixtures and trim, will be further subdivided into:

i. Water closets
ii. Urinals
iii. Lavatories
iv. Wash fountains
v. Showers
vi. Bathtubs

From this you can readily see that the list can become quite extensive whenever every possible situation is included. The important consideration here is to include the items associated with your particular project, and nothing more.

DEVELOPING ESTIMATING SKILLS

There is no short cut to developing proficiency in estimating. Most estimators go through the school of hard knocks before they can claim to be estimators. But don't be discouraged, you can learn estimating and can become quite skilled.

Most consultants usually hire persons with contracting experience because this is an important prerequisite. Inate mechanical aptitude is useful, but nothing replaces construction experience, books alone won't do it, and a college education won't do it either; however, all of the above can be very helpful.

Some consultants have progressed beyond manual estimating; they use computers. Programs are now so sophisticated that all an estimator has to do is punch in quantities for the appropriate line items and out comes the total project cost including the necessary escalation, overhead, and profit. Even with computers, all line items must be addressed.

What about facilities managers, plant engineers, and building owners and managers? How do they do their estimating? They do it carefully and thoroughly with skills developed over the years.

The first step in developing estimating skills requires an understanding of construction and installation methods. While the cost of material and equipment can be readily acquired from manuals, suppliers, or contractors, the associated labor costs depend upon variables inherent in the construction and installation process. For example, installing piping from a scaffold requires more labor than when the installation can be accomplished from floor level. Labor costs vary according to the degree of difficulty of installation. It is for this reason that contracting/construction experience is necessary.

Secondly, one must be able to *read* drawings and *interpret* specifications. For persons lacking this experience, local community colleges usually offer courses in blueprint reading and construction supervision. While home study courses are also helpful, they are usually less effective because there is no incentive for self-motivation; this depends upon the individual. Books are also available on these subjects but frequently questions go unanswered.

Back to estimating labor costs associated with specific installations. Each line item must be evaluated for degree of difficulty of installation and a factor must be included, for example:

Line Item		*Inst. Hours*		*Difficulty*	*Actual Hours*
4 in. steel pipe	×	0.10/Lf	×	1.2	.12/Lf

This means that the labor cost for the installation of a 4-inch steel pipe has a difficulty factor of 20 percent.

The estimating skills you require depend upon your responsibility within the organization of your company. In order to appropriately address the subject of developing estimating skills, it is best to start with a review of the fundamentals based on the assumption that the reader has no experience. While there are three types of estimating required in building construction, the same principles apply to all.

1. First and foremost, the ability to *read* drawings and *interpret* specifications is essential. Usually, by the time a person is promoted to a position of responsibility where estimating is required, he will have been involved in projects associated with building construction and/or renovation. Those that lack this experience must acquire it. Even for conceptual cost estimating, one must be able to visualize the building in order to develop this level of estimating.

2. Most estimators develop a history of past projects and where possible compare these estimates with actual project costs to determine performance. For those lacking actual involvement in projects to permit the development of a history, construction costs can be found in a number of trade magazines. These articles are frequently based on actual case histories, define the building or systems, and address actual costs. While the development of an estimating cost history may be laborious, it is usually well worth the effort.

3. Obtain a construction cost-estimating manual. There are several excellent manuals available and all are updated annually. The publishers of these manuals take great care to provide the latest and most updated cost data. They also provide sections of explanation that must be read and understood for optimum utilization of the cost data. Your consultant will gladly tell you which manual he uses and where you may obtain one.

4. Develop a line item checklist consistant with the type of estimate required. More serious errors are committed because line items are omitted than because of incorrect material and labor costs. The number of line items depends upon the type of estimate with a minimum for conceptual and a maximum for final estimate. Consultants generally use three different sheets when developing a final cost estimate, i.e., work sheet, estimate sheet, and summary sheet. The work sheet is actually the backup information from which the estimate is developed, and the summary sheet is used to identify costs by the various divisions of the specifications.

Worksheet

			UNIT COST		QTY. COST		
ITEM	QTY.	UNIT	MAT'L	LABOR	MAT'L	LABOR	TOTAL
Pipe, Copper Type L, 1 in.	100	LF	$ 1.50	$3.00	$150.	$300.	$450.
Valves, Gate, Soldered, 1 in.	3	EA	15.00	5.00	45.	15.	60.
						Total	$510.

The estimate sheet is developed in accordance with the various sections of the specifications. The cost associated with these sections is accumulated under the appropriate divisions of the specifications in the summary sheet. For example, all valves identified and priced on the worksheet are accumulated by type and size and listed under the section on valves on the Estimate Sheet.

Estimate Sheet

ITEM, PLUMBING	QUANTITY		AMOUNT
Valves, Copper			
Gate, 1 in.	3		$ 60.00
", $1\frac{1}{2}$ in.	5		150.00
", 2 in	5		200.00
", 3 in.	2		100.00
		Total	$510.00

After all plumbing items have been estimated, these are then totaled and listed under Division 15, Mechanical. The same method is used for the other sections of Division 15.

Summary Sheet

DIVISION 15—MECHANICAL		AMOUNT
Plumbing		$ 210,000.
Fire Protection		240,000.
HVAC		1,600,000.
Automation		150,000.
	Total	$2,200,000.

The same is done for all other divisions of the specifications. After this is all completed, the summary sheet will provide the total cost of the project.

Summary Sheet

Division 1—General Conditions		________
.		
.		
.		
.		
Division 16—Electrical		________
	Subtotal	________
Overhead and Profit	______	
	Subtotal	________
Contingency	______	
	Subtotal	________
Escalation	______	
	Subtotal	________
	Total Cost Estimate	________

Some consultants have what is known as *master specifications* that are tailored to a specific project. This technique is used to insure that all essential systems, equipment, and material are included in the final specifications. This specification is then used for estimating.

In conceptual cost estimating, plumbing was one line item under mechanical work. In the preliminary estimate, plumbing was expanded to over 10 items covering major items. In the final estimate, plumbing was further subdivided to include, in addition to common fixtures and major pieces of equipment, all specialty items necessary to provide a complete plumbing installation. The total number of line items now exceed 50.

Since the plumbing conceptual design usually contains appropriate riser diagrams and a description of the plumbing design concept only, it would be totally impractical to attempt to *take-off* such items as hangers, supports, valves, and other pipe-associated devices. Whereas at the conclusion of final design, when the entire plumbing system is fully documented on drawings and in specifications, it is essential that every identifiable item be listed.

Some consultants develop all three estimating checklists as were addressed in this chapter and are recommended as the best procedure to follow.

Developing estimating skills requires a knowledge of certain basic fundamentals. You should develop a *feel* for building square-foot costs and percentage-of-total costs for the various major trades for the types of construction you are concerned with.

The following lists the actual construction costs a school of veterinary medicine consisting of 110,000 square feet completed in 1982:

LINE ITEM	TOTAL COST	%-OF-TOTAL	PER SQ. FT. COST
General Construction	$ 6.6 Million	52%	$ 60.0
HVAC	2.9 "	23%	26.4
Electrical	1.4 "	11%	12.7
Plumbing	1.7 "	14%	15.4
Total	$12.6 Million	100%	$114.5

Next, develop unit costs for systems, equipment, and materials of construction. This should include costs of such items as: cubic yards of concrete inplace; tons of steel erected; square feet of block, brick, etc., walls; square feet of ceilings; floor covering; painting; pounds of sheet metal duct installed; linear feet of pipe; and so on.

Finally, obtain or develop a source of information on specific items costs.

You will find that you will soon develop the necessary estimating skills.

SUMMARY

Remember, an estimate is just that, an estimate and nothing more, even when submitted by a contractor as his bid proposal for a construction project. The only difference is that the contractor wagers that he can complete all the work required for that cost or less. There are times a contractor will spend more money that he estimated, and if that difference is too great or it happens too frequently, the contractor will go broke.

The consequence of your error is sometimes just as severe—you may be fired. For that reason, you should exercise just as much care as the contractor. Depending on your experience level, some or all of the following could serve you well.

1. Learn to read drawings.
2. Learn to interpret specifications.
3. Become familiar with construction materials and methods.
4. Know how buildings are constructed.
5. Understand the different types of estimates and be aware of the degree of accuracy of each type.
6. Develop a cost estimating checklist for the type of estimate required.
7. Develop an actual construction/installation cost history.
8. Update this history at least annually.
9. Obtain a construction cost data manual.
10. Get to know local vendors, suppliers, and contractors.
11. In the event you lack the experience, or the magnitude of the consequences are beyond your expertise, seek professional assistance. Many consultants provide this service and some specialize in estimating. Knowing when you are *in over your head* is more important than knowing *how* to estimate.

You can learn how to estimate, but it takes effort and diligence.

7. Remodeling Existing Buildings

Remodeling can be a frustrating, even traumatic, experience, particularly when construction costs exceed budget and the targeted completion date slowly slips by. Yet, with the escalation of land values and inflated construction costs, more and more building owners and facilities managers are turning to remodeling. Similarly, in this highly mobile society, tenants relocate frequently. Further, constant growth creates never-ending tenant expansion and this too requires more remodeling.

A recent article in a major magazine conducted a *modernization* survey, then stated continued involvement in *renovation,* and in the very next paragraph mentioned *remodeling* activities. Unfortunately, the construction industry, A/E firms, facilities and building managers, and even the press frequently use various terms to describe work done in, on, or to existing buildings.

While these various terms generally mean the same thing, there are slight variations among most. Common usage usually defines the work involved, for example, historic buildings are restored, old buildings are remodeled or modernized, while others are rehabilitated, and so on. Maybe dictionary definitions may be of some help:

Remodel	—to make new; to restore to former state
Modernize	—to make modern in style or usage
Rehabilitate	—to restore to a former capacity or efficiency
Restore	—to put back into a former or original state
Rebuild	—to make extensive repairs
Renovate	—to make new

As you can see, even this does not help too much. In order to eliminate confusion, this chapter will address all work done in, on, or to existing buildings as remodeling work. It could be extensive and involve both the interior and exterior, or it could consist of merely relocating interior partitions to suit new requirements.

REASONS FOR REMODELING

The reason for remodeling depends frequently to a great degree upon the type of work involved, and of course, economics. Actually, the reasons are as varied as the scopes of projects. One major corporation, after having purchased land in a suburban location for its new corporate headquarters, acquiesced to employees' wishes and remodeled its existing downtown property. This project, consisting of 400,000 square feet was completed for $50 per square foot, which, considering $75 to $90 for new building construction costs, yielded a significant savings. This remodeling project was considered quite an accomplishment. The work took nearly four years to complete while the employees were in the building trying to conduct business as usual. A project of this scope requires considerable preplanning and efficient execution.

Facilities managers and building owners have found that the cost for remodeling existing buildings is generally far less than the cost of new building construction. One word of caution—no major work on any building should be contemplated unless the building has a sound structural system.

There are numerous reasons for remodeling and each has its own unique decision-making and implementation process.

REASONS FOR REMODELING

1. Tenant expansion
2. New-tenant requirement
3. Tenant rearrangement
4. Refurbishment
5. Upgrading interiors
6. Replacement for obsolescence
7. Esthetic enhancement
8. Environment improvement
9. Energy conservation
10. Restoration

The reason for remodeling usually sets the tone of the project. In cases where the building was constructed or designed prior to 1973, energy conservation may be the primary reason. These buildings may be energy intensive, may lack adequate insulation, or may have too much exterior glass of ineffective quality. In rental buildings the reasons may be to suit new or existing tenant requirements.

Frequently, the primary reason for remodeling has a significant impact on other parts of the building, for example, changing an HVAC air-distribution system requires removal of ceiling and lights. While the primary reason may have been HVAC energy conservation, the installation of new energy efficient lighting fixtures may well be economically justifiable at the same time.

Another example is the exterior skin of an older building. This fenestration may be a maintenance problem, allow heat to escape in the winter, and permit excessive solar heat gain in the summer. Merely replacing the glass or windows may not provide optimum benefits. It may be well worth the added cost to install an entirely new skin to the building for aesthetic enhancement projecting a modern image. From this it soon becomes obvious that while the primary reason for remodeling may have a significant impact on the total project scope, a thorough evaluation is usually required for all but the simplest remodeling work.

The following is a list of primary reasons for remodeling and probable secondary impact:

1. Minor partition rearrangement to suit tenant or owner need. A stand-alone project usually has no secondary impact.

2. Major partition rearrangement or entire floor remodeling. This work, particularly in older buildings, usually affords the opportunity to replace or upgrade floor coverings, replace ceilings for practical or aesthetic reasons, and replace lighting with new high-efficiency fixtures and lamps for energy conservation. Where ceilings are removed, a new and improved HVAC air-distribution system should be evaluated. For entire floor remodeling work, on floors without occupants, exterior glass and window replacement should be evaluated, lobby enhancement may be necessary, and this should include such items as lighting, ceilings, floor and wall coverings, elevator doors, graphics, permanent fixtures for receptionists.

3. Energy conservation. This could include lighting improvement and power distribution system, or HVAC upgrading replacing older constant volume, terminal reheat air distribution with new variable-air-volume system and may include refrigeration equipment replacement. The amount of secondary work depends upon the scope of the primary work. This could involve demolition of the entire interior of the floor space.

4. Entire building remodeling. This is usually the most severe form of remodeling and could involve everything except the basic structural system. In some cases, the exterior walls are left intact and only the windows are replaced. Sometimes the exterior walls are modified to accept a new exterior skin that includes new windows.

5. Restoration. This is unique remodeling work associated with historic buildings. It requires the talents of experts in this work and seldom has economic justification as a prime consideration.

Regardless of the scope of the project, they all require to a lesser or greater extent a certain amount of planning, and the degree of success of the project depends upon the effectiveness of the planning program.

PLANNING REMODELING

This is actually a decision-making process. Regardless of management's role in remodeling, the process remains a constant challenge, almost like picking your way through a maze of choices with forks at every turn. Some decisions are simple and some are very complex. Suppose you own the building and the user occupant (your company) needs space rearrangement to accommodate a new management concept. The user may have done its own space planning and your maintenance personnel can accomplish the work; simple, no significant decision is required. At the other extreme, you have outgrown the building. Now what do you do? You could construct a new building and move, or you could purchase an existing building, remodel it, and move, or you could add onto your existing building to accommodate growth and remodel the existing space. Regardless of the scope of a remodeling project, a simple step-by-step procedure can be of value.

1. Remodeling Procedure
   a. Develop project scope
   b. Establish planning and design responsibility
   c. Establish budget limitation for control
   d. Develop project schedule noting major milestones

e. Select consultant if required
f. Determine project administration responsibility
g. Evaluate project results

Before the decision-making process can be addressed, the project must be defined in as much detail as necessary. This needn't be done in highly technical or legal language, but it must be written succinctly. Remember, the better the definition of a project the less opportunity for misunderstanding or confusion during later stages. After the project scope is completed, all responsible parties should be given copies for review and approval. No remodeling work should be attempted until there is agreement on the scope.

a. Project Scope

Project identification ______________________________
Address ______________________________
Project no. ______________________ Date __________
Owner's representative ______________________ Phone __________

Consultant ______________________________
Address ______________________________
Project manager ______________________ Phone __________
Scope of consulting services ______________________________
____

In this case, a decision is required and it should be made on the basis of in-house capability and availability

b. Responsibility for Planning and Design

________ In-house
________ Consultant

Should the decision be made to use outside consulting services for the remodeling project, please refer to chapter 5 for the necessary information on selecting design consultants.

It is important to develop some level of budget limitation as early as possible. This could be a simple square-foot cost estimate or it could be an upper limit established by a company executive. Once more accurate costs estimates are developed, these should be compared with the budget limitation in order to provide the necessary control and to eliminate unforeseen cost overruns.

c. Budget Control

1. Budget limitation ______________________ Date __________
2. Schematic estimate ______________________ Date __________
3. Preliminary estimate ______________________ Date __________
4. Final estimate ______________________ Date __________

5. Contractor bid ______________________ Date __________
6. Final cost ______________________ Date __________

Project description ______________________________________

____

____

____

Milestone completion dates:

Planning	________
Design	________
Construction start	________
Construction completion	________
Occupancy completion	________

Budget limitation ________

Completed by ______________________ Date __________

Complete the project scope to the extent that information is available. For example, the consultant may not have been selected, then naturally this remains incomplete until that decision has been made. The same applies to milestone dates, and budget limitation.

Next, it must be determined who will be responsible for the planning phase. This could consist of a simple floor plan sketch showing the area of construction and the work to be done, or it could be complete contract drawings and specification as developed by an A/E firm.

Time control is as important as financial control. Remember the old saying, *time is money*. This project schedule should be as simple or complex as required by the scope of the project. On larger projects that could extend a year or more from planning to occupancy, the schedule should identify each significant milestone. This is necessary in order to adequately track the project and whenever a milestone is missed, for any reason, appropriate action can be taken to correct the slippage or to adjust the schedule reflecting the slippage.

This schedule can be developed in the straightforward manner of allotting appropriate time to each phase then progressing from an established start date to occupancy. Sometimes the occupancy date is fixed as the result of a lease expiration, etc. Then the schedule must be developed in the reverse manner.

Once scheduled dates have been established, these should not be changed, instead the actual dates should be listed in the adjacent column. This will show at a glance the time status of the project.

d. Project Schedule

Milestones	*Scheduled*	*Actual*
1. Start date	________	________
2. Planning completion	________	________
3. Preliminary design	________	________
4. Contract documentation	________	________

Milestones	*Scheduled*	*Actual*
5. Award construction contract	________	________
6. Start construction	________	________
7. Complete construction	________	________
8. Start occupancy	________	________
9. Complete occupancy	________	________

The selection of the design consultant may be in the sequence listed or it may be addressed after b. Responsibility for Planning and Design. In those cases where the project responsibility is assigned to a company executive whose specialty is in a field other than technical, it is wise to select the design consultant at the very earliest opportunity. In other cases, where the in-house staff has the necessary technical expertise but either lacks adequate time for planning and design or feels that the scope of the project is beyond their limitations, this is the time to select the consultant.

The basic criteria for the selection of an architect, engineer, construction manager or, for that matter, contractor or subcontractor are the same. Choose only from those who have experience in your specific type of project and whose size and dollar volume matches the scope of your project.

Some facilities managers and building owners have a list of consultants that have performed satisfactorily for them in the past; others know of firms whose past performance has been satisfactory; while some have neither first- nor second-hand information on which to base their selection. Regardless of the situation, a simple checklist completed by the firm or firms under consideration can be a valuable aid in making a final selection.

e. Select Consultant

Most A/E firms offer a broad range of services either as part of their standard services, or ones that can be tailored to your needs. For the firms to properly assist and serve you, the range of services must match your needs.

These services include, but are not limited to, the following:

Schematic Design Phase. Consult with the owner to ascertain requirements for the project; prepare a preliminary study and provide a report consisting of suitable schematic drawings, layouts, and/or flow diagrams; prepare a budget construction cost estimate.

Design Development Phase (preliminary). Prepare preliminary documents consisting of design criteria, drawings, and outline specifications; establish scope of project; prepare preliminary estimate based upon designs established to this point.

Construction Contract Documents Phase. Prepare the required contract documents, completed drawings, and technical specifications sufficiently detailed so that the owner may obtain competitive construction contract bids.

Construction Phase. Observe the progress and quality of the construction work; endeavor to guard against apparent defects and deficiencies; review contractor's request for payment; review shop drawings, samples, and other submissions of the contractor.

The next major decision for the building manager is to determine how he wants to

CONSULTING SERVICES CHECKLIST

Firm Name: ______________________________

Address: ______________________________

Check Appropriate __ , List Number of Personnel ________

__ Architects ______

__ Electrical Engineers ______

__ Mechanical Engineers ______

__ Structural Engineers ______

__ Draftsmen ______

__ Designers ______

__ Estimators ______

Consultants (Available to Firm)

__ Kitchen ______

__ Accoustical ______

List Number for each Range of Construction Costs (Past Year)

__ Leaa than $100,000

__ $100,000 to $500,000

__ $500,000 to $1.0 million

__ $1.0 million to $5.0 million

__ More than $5.0 million

Check Appropriate __ , For Design Experience

__ Office Buildings, Industrial Parks

__ Industrial Buildings, Manufacturing Plants

__ Health Care Facilities

__ Educational Facilities

Figure 7-1. Consulting Services Checklist.

execute the construction work. This depends, to a great extent, upon the size and complexity of the project and the type of construction contract anticipated.

While construction management firms have been in existence for many years, their talents and contribution to design and construction have just begun to be appreciated and utilized.

The range of services offered by some of these firms could be as broad and encompassing as those offered by both the A/E and general contractor, with the exception of contract documents (drawings and specifications). The primary advantage of a CM is that he serves the owner, usually for a fixed fee.

While some CMs offer pre-construction planning services, A/Es feel that they are the experts in this area. Owners, they feel, should not be required to spend additional dollars on the phase.

Some construction managers do not offer these services, these CMs usually come out of the construction industry and their orientation is to construction only.

The following is a list of possible services offered by a construction management firm. It is offered here only as a guide.

Planning and Design Phase

- Preconstruction planning.
- Budget preparation and updating.
- Procurement of long-lead items.
- Prepare CPM schedule.
- Design review meeting.

Contract Procurement

- Assembly of bid packages.
- Screening of bidders.
- Bid solicitation and analysis.
- Preparation of contract documents.

Construction Phase

- Contract Administration
  —Approval of invoices and processing payments.
  —Negotiation of change orders.
  —Processing shop drawings and samples to A/E.
  —Review as-built conditions for A/E processing.
- Project Site Administration
  —Field supervision.
  —Coordination of meetings.
  —Inspect for contract compliance.
  —Interpretation of documents.
  —Record keeping.
  —Approval of work force account.
- Project Control
  —Preparation of weekly work schedules.
  —Coordination of trades.
  —Progress monitoring, update CPM.
  —Cost control and reporting.
- Occupancy
  —Coordination of owner furnished items.
  —Startup of building systems.
  —Call backs during warranty periods.

f. Evaluate Results

Probably the most difficult aspect of planning remodeling is the evaluation of results. This is complicated by the numerous intangible features of many projects. A remodeler's life would be greatly simplified if every portion of his project could be evaluated economically. For example, as in the case of manufacturing equipment where new automated machines replace manpower, the return on investment can be projected rather accurately and the decision is made simply on the basis of the acceptability of the length of the ROI. But what about the case where a significant portion of the work is aesthetic enhancement and employee environment improvement?

The results of remodeling can be best evaluated using two types of checklists. The first covers the work planned for economic reasons and is included to serve as a guide noting the more common items. Any other work accomplished for economic reasons should be added to the list.

Economic Evaluation Checklist

Description of Work	*Installation Cost*	*Annual Savings*	*ROI*
Lighting	________	________	________
Demand control	________	________	________
HVAC			
Refrigeration	________	________	________
Air distribution	________	________	________
Reduced ventilation	________	________	________
Hi-eff motors	________	________	________
Pumps	________	________	________
Automation	________	________	________
Energy management	________	________	________
Insulation	________	________	________
Boilers	________	________	________
Building insulation	________	________	________
Reduced infiltration	________	________	________
Roofing	________	________	________
Window retrofit	________	________	________

There are instances where equipment breakdown or deterioration with age occur. There are usually two options in this case; the equipment can be repaired (restored) or replaced. The best solution to this problem is to determine the two costs involved and consult with the accounting department. Experience indicates that the least cost is not necessarily the best solution. The accountant may prefer repair to replacement, as that cost is considered an operating cost deductable during the year it is incurred. New equipment must be depreciated over the life of the equipment. This choice frequently is not yours.

Aesthetic enhancement work is best evaluated on the quality of the work. While this evaluation is best accomplished through the use of a checklist. (Only a format is shown.)

Aesthetic Evaluation Checklist

Description of Work	*Quality of Work*
________	________
________	________

After the specific items of work are listed, insert the appropriate letter to indicate the quality of work. P–poor, F–fair, S–satisfactory, G–good, and E–excellent.

SPACE UTILIZATION

An essential consideration of almost all remodeling projects is space utilization. This is particularly true in cases where any significant rearrangement is planned. During this phase of a remodeling project, the work environment should be addressed. It could consist of aesthetic enhancement or it could be a change from *traditional* office space utilization to the *landscaping* technique.

Work environment is an all-encompassing phrase that probably has different meanings to different people. Here, it is used within the context of office space utilization.

Space utilization, in turn, is also a many-faceted phrase that frequently conjures up a whole myriad of connotations—"Efficiency" usually heads the list. Finally, "efficient use of space" frequently implies economy.

Yet, economic considerations should never be the sole criteria governing space utilization. Office space utilization should be based upon need. And, with optimum space utilization in satisfying these needs, economy can be the end result.

In order to arrive at this optimum use of space, a simple, step-by-step space planning procedure is advocated. The following approach is one that has been used successfully over the years and can serve as a valuable tool for anyone, regardless of their space planning experience.

An eight-step planning procedure is recommended:

- Develop an organization chart for the company, division or department involved. See Figure 2–1.
- Develop an office space standard that will be used as the criteria for space utilization. See Figure 2–2.
- Complete a projected space requirement sheet to determine the amount of space needed. See Figure 2–7.
- Develop a functional chart to determine the necessary grouping of functions.
- Draw a block diagram to show the optimum placement of these various functions.
- Establish a standard by which space utilization will be carried out.
- Determine the placement of ceiling-high partitioned work stations and enclosures (offices, conference rooms, etc.).
- Develop the final space layout.

There are generally no short cuts to space planning. In fact, some of the larger space planning organizations use a far more complex and detailed procedure. However, these eight steps offer sufficient tools for efficient and professional planning.

However, a procedural discussion should be expanded to include all the necessary considerations toward eliminating error, reducing confusion and increasing planning effectiveness.

The first step in this procedure is to *develop an organization chart*. In many companies, this is an ongoing function of the personnel department. They develop the initial chart and update it whenever changes are significant enough to warrant a new one.

While there are samples of more-or-less typical charts available, one may be tailored to reflect a particular company's organization.

In a large company, one organization chart may be necessary for every different department; and one chart used to show these departments in their respective positions relative to one another.

When completing the chart, it is important that every different level position be shown in the appropriate row. Just because a supervisor, a manager, and a vice-president all report to the president doesn't mean that they all have the same level of authority and responsibility—or require the same degree of office facilitation.

Secondly, it is essential that these various positions be assigned different levels, since levels are used to determine the type and size of space allotted in accordance with space standards. Further, these different levels of supervision and management have different needs as far as overall space and type and size of furniture.

It is considered good management practice to limit the number necessary to perform the appropriate functions of authority and responsibility. Unfortunately, the space planner usually has no voice in determining the organization chart. This is normally company policy established to satisfy company needs as determined by the top executives.

Establishing office space standards is important for several reasons. First, it establishes the amount of space assigned to the various position levels and, second, it eliminates hassling about the size and type of individual offices. Once persons are slotted on the organization chart, the size and type of their work stations are also determined.

Armed with the organization chart and the office space standards, it now becomes a relatively simple matter to *complete projected space requirements.*

A projected space requirements sheet could consist of space group designations by executive need, giving the number in each executive group or function, space per group and total space in square feet. Miscellaneous areas such as conference rooms, storage rooms, etc., should be listed in the same manner. Total space necessary plus traffic area space (20% of total) are compared against total usable space.

"Usable" space is that space where carpet can be laid and furniture placed. It is significantly different from "rentable" space and "gross" space. In other words, usable space *excludes* such areas as toilet rooms, mechanical rooms, stairwells, elevator shafts, elevator lobbies, etc.

In a typical case where an entire floor is leased, "gross" space is the entire area of the floor usually measured from the exterior wall glass line to exterior wall glass line. "Rentable" space is that area in square feet upon which the rent is based.

In the typical case mentioned above, rentable space is frequently about 5 to 20 percent above the usable space to account for those areas not used exclusively by the occupant of the floor. It usually excludes those areas that penetrate the floor, for example, stairwells, elevator shafts, HVAC ducts, etc.

Understanding these differences is important because if you need 10,000 square feet of usable space, renting 10,000 square feet usually will not be sufficient. Approach your space project with this understanding in mind.

FUNCTIONAL CHART

Code	Department	Area Sf Ft	Location	Floor	Adjacency
A	Accounting	8,000		2	E & H
B	Personnel	3,000	Front Entrance	1	C
C	Purchasing	2,700	Front Entrance	1	B
D	Sales	5,500		2	E
E	Marketing	6,200		2	D
F	Facilities	1,800		1	C & G
G	Shipping & Receiving	2,000	Loading Dock	1	-
H	Stock Room	6,000	Loading Dock	1	G
I	Corporate Staff	6,000		2	-
J	Lobby	200	Front Entrance	1	-

Figure 7-2. Functional Chart.

Develop a functional chart so that you can determine the necessary groupings by function and the size of the space in square feet.

Some space planners skip this step because they feel that they can estimate the size of these functions by reviewing the Organization Chart and the Projected space Requirements Sheet.

This practice is not recommended. There are no short cuts to space planning. Develop the functional chart!

This functional chart provides the basis for developing *a block diagram.*

This block diagram plays an important role in the total space planning program. It is the first step in the preparation of the final space design drawing because it permits the proper placement of the various functions depending upon their interdependent roles.

Second, a number of different functional placements can be quickly developed, and critiqued, until the optimum locations are developed.

Further, since this is usually performed in free-hand, valuable time is saved over going directly to the hard line, final design drawing.

You are now at the cross-roads of office space planning. At this point a choice will have to be made between the "traditional" office space utilization and the "landscaping" office techniques. This presentation will not address itself to the merits of one method over the other, as there have been many excellent pieces written over the years covering the advantages and disadvantages of the two methods.

This is geared so that the entire procedure is applicable in either case. In the event you have elected to go the landscaping route, review the office space standards introduced above and disregard references to "type of space." The "size, sq. ft." for the various space group designation will, in general, hold as defined.

An important consideration in an office space planning program is the *placement of the ceiling-high enclosures.*

One approach is to place all such offices, conference rooms, etc., in the interior areas so that maximum advantage is taken of the window walls providing natural light to the space.

The other approach is to place all ceiling-high offices along the window space. Each method has its own unique advantages, the choice depends entirely on where the emphasis is placed.

Frequently, the decision is based upon past experience and preference. Some say that executive offices should be placed along the windows so that these people will have the benefit of the spacious atmosphere created by the exterior glass wall in their offices.

The recommendation is a combination of the two; some offices along the windows and some along the interior. Specifically, conference rooms, storage rooms, etc., should always be placed in interior spaces. Naturally, the president's office should be so located that he has the best view from his normal seating position.

The first step in developing the *final space layout* is to obtain an architectural drawing of the building or space involved. This drawing should show all permanent walls, enclosures, etc.

Overlay this drawing with tracing paper so that preliminary free-hand layout plans can be developed. Since this tracing paper is economical, it can be used to develop several different plans until one is found that best suits varying needs.

If your department has the appropriate drafting talents, the next step is to develop the hard-line drawing. Otherwise, this part of the work can be subcontracted to the proper firm with the requisite talents.

ESTIMATING REMODELING COSTS

Estimating remodeling costs is somewhat different from estimating new construction. While the actual method and techniques are the same, the process is different. This is so because in most cases the work is limited to interior space and in almost every case demolition is involved.

Demolition costs are usually most difficult to estimate for several reasons. There is often unseen work involved that is overlooked, the cost of trash removal from the particular floor to a central location at the building is difficult to estimate, and the cost of trash removal from the building to an authorized dump site depends mainly on haulage mileage and dumping charges which are frequently unknown at the time of estimating.

The same technique is used here as with any other type of estimating. The first step is to develop the appropriate checklists including all possible items of work. Since there are actually two distinct phases of work involved each should be addressed individually.

Demolition Estimating Checklist

Description of Work	*Length, Area, or Quantity*	*Unit Cost*	*Total Cost*
Partitions			
Block	______	______	______
Dry wall	______	______	______
Bank screen	______	______	______

Description of Work	*Length, Area, or Quantity*	*Unit Cost*	*Total Cost*
Ceiling	______	______	______
Lights	______	______	______
Floor covering	______	______	______
Drapery	______	______	______
Blinds	______	______	______
Receptacles	______	______	______
Switches	______	______	______
Phone outlets	______	______	______
TWX outlets	______	______	______
VDT outlets	______	______	______
Xerox outlets	______	______	______
Electrical panels	______	______	______
HVAC			
Terminal Units	______	______	______
Duct work	______	______	______
Exhaust fans	______	______	______
D. X. units	______	______	______
Piping	______	______	______
Sprinklers			
Heads	______	______	______
Risers	______	______	______
Hose cabinets	______	______	______
Fire extinguisher	______	______	______
Plumbing			
Water closets	______	______	______
Urinals	______	______	______
Lavatories	______	______	______
Partitions	______	______	______
Water coolers	______	______	______
Elevators	______	______	______
Kitchen equipment	______	______	______
Trash removal			______
Clean-up			______
Patch and repair			______
Haulage			______
Total cost			______

Again, as with past checklists, this is intended to serve as a guide only, and cannot of necessity include every item of work for all projects. Use it as a guide, add or delete as required for the remodeling work at hand. A similar checklist must be developed for installation work. Since the format has been shown for the demolition, only those line items will be listed that may be required and are not included above.

Supplementary Installation Checklist

Description of Work	*Length, Area, or Quantity*	*Unit Cost*	*Total Cost*
Partitions			
Wall covering	________	________	________
Paint	________	________	________
Base 4″ vinyl	________	________	________
Hardware	________	________	________
Glass	________	________	________
HVAC			
Smoke detectors	________	________	________
Smoke dampers	________	________	________
Fire dampers	________	________	________
Smoke purge	________	________	________
Stairwell pressure	________	________	________
Fire alarm system	________	________	________
Security system	________	________	________
Communication	________	________	________
Art work			
Graphics	________	________	________
Paintings	________	________	________
Sculpture	________	________	________
Plantings	________	________	________
Specials			
Chalk boards	________	________	________
Tack boards	________	________	________
Paneling	________	________	________
Built-ins	________	________	________

In cases where the exterior of the building is involved in remodeling, that must be added in both demolition and new construction.

The techniques of estimating are described in chapter 6 and will not be addressed here.

SUMMARY

Even the simplest remodeling work requires the same attention as new work does.

1. Define the scope of the project in sufficient detail commensurate with the type and amount of work involved.
2. Determine who will be responsible for planning and design.
3. If you lack in-house talent or time, select an appropriate consultant.
4. Establish a project schedule noting all essential milestones. This is needed to

determine if and where the project is slipping so that appropriate corrective action can be taken.

5. Always establish budget limitation.
6. Develop the cost estimate as early as possible, and in enough detail so that costs can be monitored.
7. Evaluate results for both economics and aesthetics.

8. Construction Administration

One of the most important functions of the construction administration phase usually takes place soon after the construction contract is awarded and before construction starts. It is the ground-breaking ceremony. Here the owner or his representative, armed with a gold-plated shovel, breaks ground in the presence of the contractor, consultant, and, depending upon the importance of the project, local, state and/or national dignitaries. Frequently, short speeches are given by the major participants, and when the event is important enough it may receive press and/or TV coverage.

At the other end of the project, an equally important function is performed, and this is the dedication ceremony. Here again, appropriate individuals make speeches giving credit to project participants, usually high-level executives. Sometimes, unfortunately, the people who held day-to-day responsibility for construction are not acknowledged. These are the ones who take a set of contract documents that frequently are not perfect, assemble material and equipment, and erect a building under sometimes adverse conditions, on a site that may not be best suited to the project. The period between ground breaking and dedication is fraught with seemingly insurmountable obstacles, the work may be hampered by strikes, delayed by unfavorable weather conditions, and complicated by equipment delivery delays. But it is the period that yields great satisfaction of accomplishment to a dedicated, though often unheralded few who somehow get the work finished.

The construction phase begins with the award of the construction contract and terminates when the final payment is made to the contractor by the owner. The consultant considers his obligation fulfilled, when the owner accepts the final certificate for payment. When this final payment is received by the contractor, the project is closed.

BIDDING PERIOD

In most instances, the relationship between owner and contractor, and consultant and contractor begins after the completion of contract documents during the bidding period. During the final design phase, the consultant prepares a bidder's list based on past experience, qualification statements from bidders, and/or input from the owner. The consultant, through the specifications, notifies all bidders when and where they may pick up contract documents, establishes the time and place for a prebid meeting, describes the bidding procedure, and notifies bidders of delivery time and place for all bids.

During the bidding period, owners have certain responsibilities. Usually the owner advertises for bids from contractors on the previously prepared list. The owner receives, tabulates, and with the assistance of the consultant (sometimes the consultant performs these services), analyzes the bids. The owner issues a letter of intent to the selected contractor, then prepares and executes the construction contract. At this point, it is to be hoped that there is full agreement on the work to be performed and the cost of the project. The owner should also attend the prebid meeting.

The prebid meeting serves a useful function in all but the smallest projects. This meeting is held at about the midpoint of the bidding period, usually after the contractors have had several weeks to review the drawings and specifications. The intent of this meeting, attended by all bidders, sometimes the subcontractors are included, the owner, and the consultant, is to clear up misunderstandings and to rectify discrepencies. In those cases where the consultant did not have sufficient time to complete his in-house checking of the contract documents prior to issue, this work is performed during the bidding period. This information is conveyed to the contractors in the form of an addendum, similar to the questions, comments, answers, and clarifications noted during the prebid meeting. The contractors are usually allowed at least one week to incorporate the addendum(s) into their bids as all addendum(s) become part of the contract and the cost of all work is included in the bid.

CONSTRUCTION CONTRACT DOCUMENTATION PHASE

As was stated in chapter 5, complete drawings and specifications do not exist. For this reason, there will be clarifications, modifications, and revisions, and in many cases the owner will make changes to the contract to suit revised requirements. Since this is an accepted fact, something must be done during the contract documentation phase to anticipate these changes to simplify the construction administration phase.

Further, well-prepared specifications will require the contractor to provide other appropriate documents that are necessary so that the owner and consultant can adequately monitor construction progress and evaluate progress payment requests. These may include some or all of the following:

- Changes in the work
- Construction progress schedule
- Shop drawing submission schedule
- Schedule of values
- Manpower loading
- Cash flow chart

Changes in the work

These changes, additions, or deletions must be addressed as part of the construction contract. Since most changes in work will cause a change of contract price, it is essential that the method of charges for this work be clearly established. In order to simplify administration of changes and payment for this work, sections covering Unit Price Work and Nonunit Price Work are included in the specifications.

Unit Price Work. The unit price schedule should include all items that can normally be expected to be added to or deleted from the contract. In addition to the schedule of prices, certain clarification statements are required to define what is included in the work and what conditions control the quality of the material, installation, and workmanship.

- The contractor agrees that the owner may make changes without invalidating the contract.
- All equipment purchases and the installation methods shall be bound by the specifications.
- Unit prices shall include the cost of materials, work, layout, drafting, balancing and testing, tools, and sundries, taxes, permits, coordination, overhead and profit, supervision, and any and all other costs for the completed work in place and accepted, or omitted, as the case may be.
- All unit price work shall be satisfactorily completed and accepted by the owner and consultant.
- Unit prices shall apply throughout the life of the contract.
- The contract sum shall be adjusted accordingly.

Since it would be impossible to list all work that may be expected to be added to or deleted from every conceivable project, the following is a list of typical items showing categories, and does not include, for example, all different sizes of pipes, valves, conduit, wiring, etc. The number of different sizes of pipes, valves, etc., depends upon the individual project. The cost for add or deduct are typical prices charged at this time in a specific location of the country and are not intended to reflect costs elsewhere at a later time.

SCHEDULE OF UNIT PRICES

Item No.	*Description of Work*	*Unit*	*Add*	*Deduct*
1.	Earth excavation	CY	$ 3.50	$ 2.50
2.	Rock excavation	CY	75.00	N/A
3.	Concrete walks, including base course, in place broom finish	SY	16.00	14.00
4.	Concrete curbing in place	LF	5.50	4.50
5.	Bituminous paving including top and base course in place (parking type)	SY	16.00	14.50
6.	Sodding, in place	SF	0.30	0.20
7.	Concrete footings in place	CY	50.00	40.00
8.	Concrete walls in place	CY	55.00	45.00
9.	Concrete slab on grade in place	CY	50.00	40.00
10.	Reinforcing steel in place	LB	0.70	0.50
11.	Wire mesh in place	CSF	30.00	22.00
12.	8″ CMV wall in place	SF	3.60	3.00
13.	Drywall partition, in place	LF	18.00	14.00
14.	Vinyl asbestos tile, in place	SF	0.80	0.60
15.	Vinyl wall covering, in place	SY	8.00	7.00

Item No.	*Description of Work*	*Unit*	*Add*	*Deduct*
16.	Painting on drywall	SF	0.25	0.20
17.	Hot and cold water piping connected to fixtures and risers, including fittings, insulation and hangers—1″, copper	LF	25.00	18.00
18.	Soil and waste piping connected to fixtures and stacks, including fittings, supports, and accessories —4″, cast iron	LF	30.00	21.00
19.	Hot and cold water valves, 1″ copper	EA	35.00	23.00
20.	Floor drains, in place—4″	EA	250.00	160.00
21.	Sheet metal, rectangular, low pressure, Gal. Iron	LB	4.00	1.50
22.	Duct insulation, rigid, 1″ thick	SF	5.50	3.00
23.	HVAC piping, with insulation, fittings and hangers, 1″	LF	30.00	20.00
24.	Glass thermometer, 8″ with socket, in place	EA	160.00	100.00
25.	Space thermostats	EA	120.00	90.00
26.	Relocate space thermostat cost for pneumatic tubing, in place	EA	5.00	3.00
27.	Full weight galvanized rigid conduit complete, exposed with couplings, locknuts, threading, bending, pull boxes, supports and elbows, in place, 1″	LF	3.00	2.50
28.	Single conductor THWN copper in conduit, all splices, connections and accessories, #10	LF	0.40	0.35
29.	20 ampere, single pole switch with box, extension cover and plate, in place, sheetrock walls	EA	35.00	30.00
30.	20 ampere, duplex receptacle, complete with box, extension cover and plate, in place, sheetrock wall	EA	32.00	27.00

Nonunit Price Work. Since there will always be work that cannot be included in unit-price work, this must be covered in sufficient detail that the contractor will be adequately compensated and the owner properly protected.

- Contractor will be provided with adequately detailed drawings and sufficient description to permit an accurate estimate of the cost of the work.

- Such estimate shall include contractors' and subcontractors' overhead and profit markup where supervision is required for the work, as follows:

1. Contractors' own work—20 percent
2. Contractor supervision—10 percent
3. Subcontractors' own work—20 percent
4. Subcontractor supervision—10 percent
5. Subordinate sub's own work—20 percent
6. On work or supervision omitted—0 percent

These percentages for additional or omitted work cover entire amount applied to overhead and profit and cannot be changed during the life of the contract.

- All equipment purchased and the installation methods shall be bound by the specifications.
- The contract sum shall be adjusted accordingly.

Construction Progress Schedule

The contractor is usually instructed in the specifictions to prepare and submit to the owner a construction progress schedule within a specified time after being awarded the contract, or at a specified time before submitting his first application for progress payment. This could be in the form of a CPM (critical path method), PERT (program evaluation and review technique) or the familiar bar chart. Regardless of the format used, the progress schedule should list all major events showing their beginning, duration, and completion. Some events may be discontinuous, in that case separate start, duration, and completion must be shown for each portion of the event. While the schedule is not considered a contract document and it has no legal status, it could be used to validate or counter delay claims. It provides the owner and consultant with an instrument for monitoring construction progress and it is used to identify slippage as soon as it occurs. This should then be brought to the attention of involved parties so remedial action can be taken to bring the construction back on track.

Shop Drawing Submission Schedule

The contractor is usually instructed in the specification to prepare and submit a shop drawing submission schedule within a specified time after being awarded the contract. Since the consultant reviews and approves, or takes other appropriate action upon the contractor's shop drawings, this schedule is necessary in order that the consultant can properly evaluate the work load imposed by the required review. Once the schedule is developed and submitted, it is important for the contractor to make every effort to adhere to the schedule.

Shop drawings are an all-encompassing term that includes drawings, product data, and samples. Shop drawings illustrate portions of the work such as: equipment

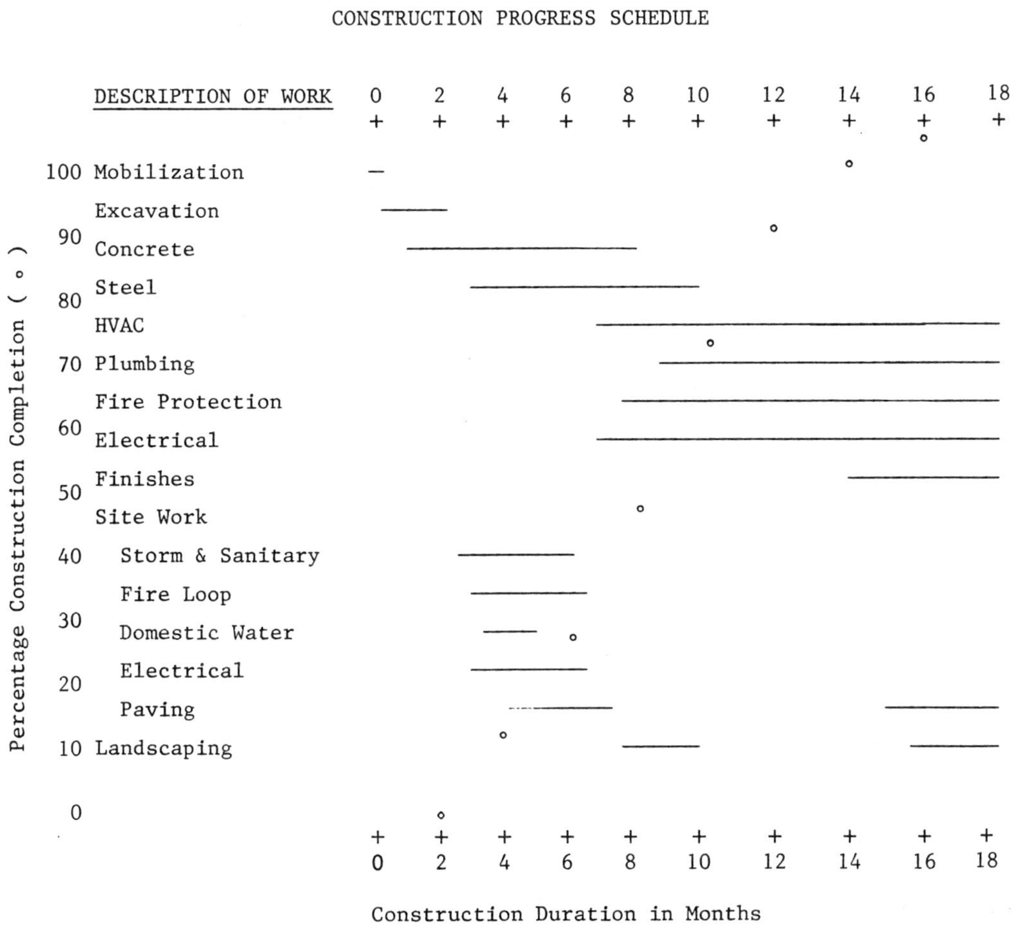

Figure 8-1. Construction Progress Schedule.

layout; concrete and reinforcement; sprinkler layout, etc. Product data are usually manufacturers' illustrations, diagrams, brochures, or instructions. Samples are usually physical samples which illustrate color, texture, and quality of materials. Some shop drawings are coordination drawings showing relationship of equipment, piping, ducting, lighting, etc.

Schedule of Values

The schedule of values is actually a construction cost breakdown. It is a requirement of the specification and the contractor is instructed to submit it before he submits his first application for payment. While not a requirement in all cases, the schedule of values should include quantities and unit prices aggregating the contract price. It should subdivide the contract price. It should subdivide the work in sufficient detail to serve as the basis for evaluation of progress payments during construction.

On a recent $8 million project, the contractor submitted a schedule of values identifying, by individual cost, 145 line items. This submission was considered satisfactory to the consultant and owner and it permitted efficient and accurate evaluation of applications for payment. It contained 14 major categories following the general outline of a typical specification, except that Mechanical was separated by trade to show HVAC, Plumbing, and Fire Protection. The following is a shortened form of the above with a cost breakdown shown for General Requirements only and noting the number of line items for the balance.

SCHEDULE OF VALUES

General Requirements		
1. Bond premium	$ 65,000.00	
2. Mobilization	35,000.00	
3. General conditions	300,000.00	
Site Work		
4. through 13		
Concrete		
14. through 21		
Reinforcing Steel		
22. through 23		
Masonry		
24. through 29		
Doors and Windows		
30. through 40		
Finishes		
41. through 55		
Specialties		
56. through 64		
Equipment		
65. through 70		
HVAC		
71. through 100		
Fire Protection		
101. through 108		
Plumbing		
109. through 119		
Electrical		
120. through 145		
Total		$8,000,000.

One consultant developed a special schedule of values form that the contractor was instructed to complete prior to submission of his first application for payment and to update for each subsequent application. This form showed the exact financial status of the project for each submission.

SCHEDULE OF VALUES
Smart Consultants
Somewhere, N.J.

Project ______________________ Project No. __________
Owner ______________________ Application No. __________
____
______________________ Application Date __________
Contractor ______________________
______________________ Application Period:
______________________ From ________ To ________

SPEC. DIV. NO.	DESCRIPTION	COST IN DOLLARS	PREVIOUS VALUE	VALUE THIS REQUEST	TOTAL TO DATE	COMP %
1	Gen. Req'mts.	$ 50,000.	$10,000.	$20,000.	$30,000.	60
2A	Earthwork	15,000.	5,000.	5,000.	10,000.	67
2C	Bitum. Paving	3,000.	0	0	0	0
2E	Seed & Sod	5,000.	0	0	0	0
2F	Landscaping	1,000.	0	0	0	0
—	—	—	—	—	—	—
16	Fire Alarms	25,000.	0	0	0	0
	Totals	$2,500,000.	$15,000.	$25,000.	$40,000.	1.6

(Dash (—) indicates omitted line items and totals include their costs)

This format not only satisfies the basic requirement for a schedule of values, it provides a running account of the percentage of completion of every line item at the time of submission of an application for progress payment. By updating the bottom line totals, the entire project expenditure can be readily determined.

A somewhat different format was used on a federally funded state project. Here, in addition to the normal schedule of values showing a total cost per line item, the form provided for and the contractor submitted number and kind of unit, cost per unit, material cost, labor, and other costs. This provided several additional means for evaluation of the request for progress payment. However, in this case the consultant was required to keep the running total separately.

DETAILED COST BREAKDOWN

Project ______________________ Contract No. __________
Address ______________________
Contractor ______________________
Address ______________________

ITEM NO.	DESCRIPTION	UNIT NO. & KIND	COST/ UNIT	MAT'L. COST	LABOR COST	TOTAL COST
1	Permits	1 ea	$1,500	$1,500	$ 0	$1,500
—	—	—	—	—	—	—
3	Security Fence	300 LF	9	1,100	1,600	2,700
—	—	—	—	—	—	—
9	Paving-Bitum.	200 SY	15	1,500	1,500	3,000
—	—	—	—	—	—	—
50	Fans-Exhaust	10 EA	800	7,200	800	8,000
			Totals			$1,500,000

As you can readily see, there are several ways that schedule of values can be requested. Determine the one that best suits your needs, or leave it up to your consultant, as he may have one that best suits his purposes because he is usually responsible for the evaluation of requests for payments and to maintain a running project status.

Manpower Loading

While this could be of value during construction to monitor the contractor's concept of the number of people required at different times for the period of construction, it is usually not a very accurate method for monitoring progress. In the first place, the contractor usually identifies this as *estimated project manpower,* and unless the contractor must submit a weekly manpower log there is no way to validate the number of people on the job at any one time. Consultants cannot be expected to count heads. This sometimes turns out to be just an interesting bell-shaped type of curve.

Cash Flow

Cash-flow information is valuable to the one who spends money. In construction, this is usually the owner, unless the project is financed, then the lending institution will have to know. The cash-flow information could be submitted either in a tabular format or as a graphic presentation. In either case, it must show a running total, cumulative by billing period, for the duration of construction.

It is interesting to note what portion of the money is consumed during different periods of a project. For example, on a recent $8 million, 16-month construction project, the following is actual cash flow, and it is typical for commercial, light industrial, and institutional projects.

Period	*% Total Consumed*
1st 4 months	11%
2d 4 months	41%
3d 4 months	35%
last 4 months	13%

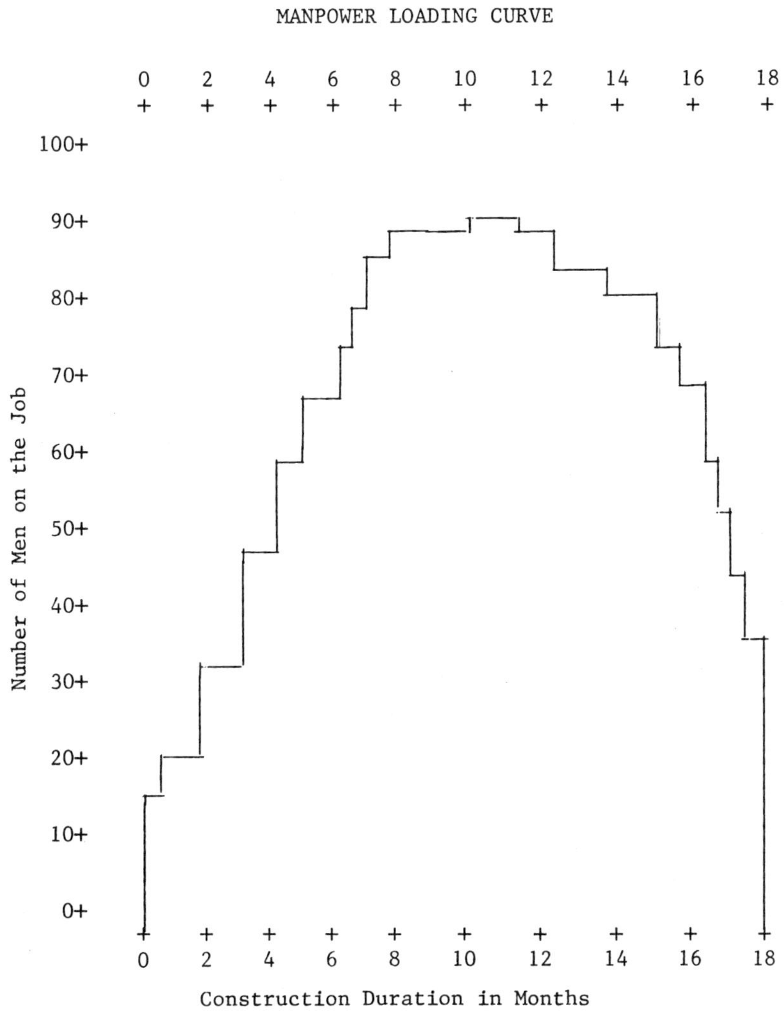

Figure 8-2. Manpower Loading Curve.

When the dollars per billing period is plotted against project time, a bell curve very similar to the one for manpower loading is developed. This is to be expected as manpower requirement slowly builds up during the early phase of construction and tapers off near the end of the project.

On a typical project, similar to the illustration above, about 25 percent of the money is consumed during the combined first and last 25 percent of the construction period, and about 75 percent is consumed during the middle 50 percent period. In the event that a contractor indicates large cash flow at the beginning of a project, he may be trying to enhance the firm's cash position. However, tread cautiously for the contractor may have legitimate, up-front, high capital investments for expensive equipment.

Most contractors provide a billing curve that indicates projected cumulative billing

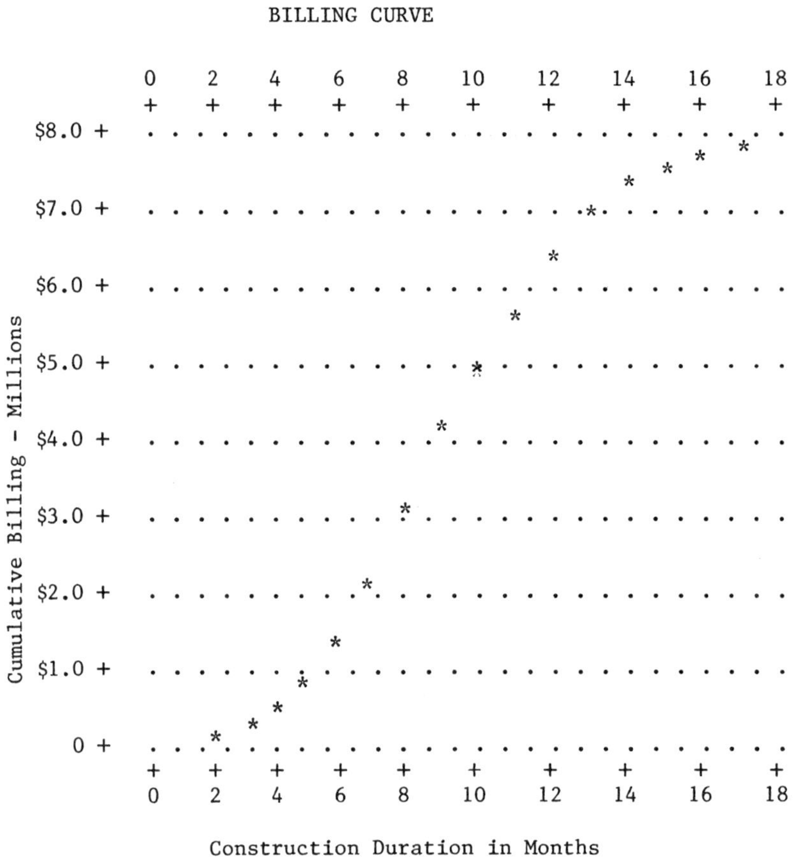

Figure 8-3. Billing Curve.

plotted against construction duration period. This should be resubmitted at each billing period and updated to reflect actual billing to date.

CONSTRUCTION ADMINISTRATION

Drawings and specifications have been completed to your satisfaction, bidding has been accomplished, the construction contract has been awarded, and the construction phase starts. Similarly, construction administration begins.

The construction administration phase is usually one of the five basic services offered by consultants. While some owners do not contract for this service, it is usually not in the best interest of the owner to omit it, and secondly, most consultants insist that this be part of their contract.

The consultant usually performs the following as part of his normal obligation or as agreed upon between owner and consultant:

1. Serves as the owner's representative.
2. Advises and consults with the owner.
3. Forwards owner's instructions to the contractor.
4. Visits the site at regular and appropriate intervals. However, the owner may,

depending upon the size and complexity of the project, contract for, under additional services, full-time, on-site representation.

5. Conducts job meetings at the site at regular intervals, usually once every two weeks, and writes minutes of the job meetings. On federally funded projects, this function is usually performed by the appropriate government representative.
6. Interprets the requirements of contract documents.
7. Endeavors to guard the owner against defects and deficiencies in the work of the contractor.
8. Issue Bulletins, Change Orders, and/or Construction Change Authorizations as required.
9. Monitors construction progress.
10. Reviews shop drawings, product data, and samples.
11. Evaluates contractor's application for progress payments.
12. Conducts special site visits at approximately 90 percent construction completion to evaluate final compliance with contract documents and notate findings on what is known as a *punch list.*

The consultant is, generally speaking, the owner's representative whether based full time at the site or conducting periodic visits to the site. However, there are limitations to the consultant's responsibility. For example the consultant:

1. Does not supervise the work.
2. Is not responsible for the contractor's means, methods, sequences, or procedures of construction.
3. Is not responsible for safety precautions.
4. Is not responsible for the contractor's failure to do the work as described in the contract documents or for any omissions.

One word of caution for the entire construction administration phase. There have been cases of severe legal consequences as the result of improper acts or lack of acts during construction. The assignment, establishment, acceptance, or avoidance of responsibilities must be performed in accordance with accepted legal practice. Numerous lawsuits have been generated during this phase of a project. Because of the probable legal liability, most owners retain the consultant who designed and developed the contract documents for the project.

Drawings and Specifications (Project Manual)

You should have reviewed the drawings and specifications several times during the design and again after final issue for bidding. If you had properly and thoroughly conveyed your general and special requirements to the consultant, the final set of contract documents should reflect these needs. The importance of your input during program planning, schematics, and design development cannot be overemphasized. As was stated elsewhere in this book, the consultant is not a mind reader. He knows your needs and desires only to the extent that you delineated them.

The contract documents are complementary in that what is called for in one is as binding as if called for in all. This forms the basis for a contract between the owner and contractor, and is generally subdivided into the agreement, the general conditions of the contract (and includes supplementary conditions which usually modifies the general conditions), technical provisions, and drawings.

The drawings are generally diagrammatical in nature and show the character and scope of the work and describe quantity of materials and the method of assembling them. Figure dimensions take precedence over scale dimensions and detail drawings take precedence over general drawings. In cases of conflicts between the several contract documents, court rulings have given precedence to the specifications over the drawings and the agreement over either.

The document that was and, in some cases, is known as The Specifications is now being referred to as the *Project Manual* and that portion of The Specification previously called Technical Provisions is now referred to as the *Specifications.* While this change in terminology has yet not been universally accepted, it is being used more frequently. For the purposes of this book, the term The Specification will continue to be used.

While it is the consultant's responsibility for interpretation of the drawings and specifications, you should be thoroughly familiar with both. You should read the specifications, word for word, from cover to cover in order to be sure that its contents truly reflect your needs and desires. The same diligence applies to the drawings to insure that the placement of equipment and the method of installation will best serve your needs in the future. Check to see that in the event of equipment failure at some future date, it can be removed and replaced.

Here are several concepts of good design used successfully in the past:

- Special knock-out panels were designed for ease of removal adjacent to HVAC chillers to provide access to the equipment.
- In the basement of a four-story building, the horizontal run of the roof drains exited the building at an elevation above the level of the adjacent retention pond. After the conductor exited the building, it was lowered to enter the pond at a point below the surface.
- The louver for ventilation of the emergency generator room was increased to a size sufficient to permit removal of the emergency generator.
- A manual jib hoist was installed on a roof so that heavy equipment could be swung out over the parapet and lowered to ground level.

The list could go on indefinitely but it is clear that these could be maintenance features and not necessarily design problems. Consultants are well aware of these conditions and usually include them in the design.

Product Specifications

The Specifications provide technical description of materials, equipment, construction systems and methods, standards, and workmanship. The Specifications also address options and substitutions for products specified.

There are generally four conditions describing specified products:

- Where products are specified by naming only one manufacturer, there is no option for substitution. This is usually a condition imposed by the owner to provide a match to existing products, or because of his personal preference. While this usually simplifies the design phase of the project, it could also increase the total cost of the project because competition has been eliminated.

• Where products are specified only by reference standard, any product meeting that standard may be used. This is common practice on federally funded projects, and is usually used elsewhere for such items as pipe, fittings, conduit, wire, hangers, steel, concrete products, etc.

• Where products are specified by naming several manufacturers, any one of the manufacturers named may be used. This has several excellent features. First, it permits competitive bidding and yet establishes a quality of product.

• Where products are specified by naming one or more manufacturers and "or equal," or "or approved equal," the contractor may submit a request for any manufacturer not specifically named. While this option provides the most incentive for competitive bidding, it also creates the greatest amount of confusion and frequently provides frustration for all concerned. The greatest problem with this option is the evaluation of what constitutes *equality*.

A well-developed specification defines when and how substitutions are to be submitted and who has final responsibility for the acceptability of the proposed substitution.

• Before bid submission—not less than 10 calendar days before bids are due. If these substitutions are acceptable, then all bidders are informed so that these may be included in their bids.

• With bid submission—these substitutions are submitted as alternates to the base bid and their cost impact on the base bid must be clearly delineated.

• After contract award—usually not later than a specified number of days after the award. This substitution is sometimes a part of the shop drawing submission list where the contractor identifies all substitutions with an asterisk or similar notation. The consultant then indicates his preliminary approval or rejection and returns the list to the contractor. This preliminary approval indicates to the contractor that he may include it for submission in accordance with conditions usually included in the specification but the consultant is not bound to approve. He will evaluate and judge the product when complete data are submitted at a later date.

The contractor must comply with the following conditions for this submittal. The request for substitution must include complete data with drawings and samples, as required, including the following:

- Quality comparison—proposed substitution versus specified product.
- Changes required in other work because of the substitution.
- Effect on construction schedule.
- Cost data—proposed substitution versus specified product.

Additionally, the contractor states that he:

- has investigated the proposed substitution and has determined that it is at least equal to the specified product;
- has waived all claims for additional costs as a result of the substitution;

- will coordinate and make other changes as required as a result of the substitution;
- will guarantee the substitution in the same manner as the specified product.

Finally, most specifications state that the consultant is the sole judge of the acceptability of the proposed substitution.

Changes During Construction

Changes prior to award of the contract are covered by the addendum(s) while changes after the contract are handled by modifications or bulletins.

This bulletin becomes part of the construction contract documents and must be handled in a professional manner. Most consultants develop a standard format that will be used throughout the life of the project. The information contained in the document must be as complete as is necessitated by change. Sometimes a one-pager will suffice, and sometimes several pages will be required plus all appropriate drawings.

SAMPLE BULLETIN FORMAT

Project ________________________ Project No. __________

____

______________________________ Date __________

Bulletin No. __________

Part 1—General

1. This work is a result of unforeseen field conditions and you are authorized to proceed. (In some cases the contractor is instructed to submit a cost proposal for review and approval before proceeding.)
2. The price change shall be determined as required by the General Conditions. A complete cost breakdown with back-up material shall be furnished with the contractor's proposal.
3. The scope of work includes ______________________

____

Part 2—General Conditions

Furnish all labor, materials, and equipment, make necessary revisions, additions, and deletions in accordance with the drawings and written material issued as part of this Bulletin.

Part 3—Description of the Work

Drawing No.	*Description of Change*
M-1	______________________
	______________________
S-1	______________________
	______________________

Part 4—Enclosures

M-1, Dated ______________

S-1, Dates ______________

Signature of Consultant ______________________

Distribution:

Once the contractor's cost proposal is received, it is evaluated, and approved or rejected. If approved, a formal Change Order is issued to the contractor. The appropriate cost changes become part of the contract.

Another document used frequently is the Field Order. This is used generally for clarifications that do not involve contract price changes.

Finally, there is the Construction Change Authorization. This document is issued to expedite construction and avoid or minimize delays and is used where the work may affect contract sum or contract time. It authorizes the contractor to proceed immediately and after the cost has been determined, a Change Order is issued to reflect the change to the contract price.

Record Keeping

A most important part of the construction phase is record keeping. This applies equally to the contractor, consultant, and owner. Since the contractor usually keeps all appropriate records deemed necessary for his purposes, you should see to it that all records that may be required are available.

In the event that you did not contract for full-time on-site representation by the consultant, the task of record keeping falls squarely on your shoulders.

The contractor is usually required, by the specifications, to maintain certain records. Most specifications require the contractor, under Project Record Documents, to maintain at the job site, one copy each of the following:

1. Drawings
2. Specifications
3. Addenda
4. Bulletins
5. Field orders
6. Construction change authorization
7. Change orders
8. Written instructions
9. Shop drawings, product data, and samples
10. Field test records
11. Construction photographs
12. Daily manpower log
13. Daily progress and status notations

In addition, the contractor is instructed to provide files and racks for the storage of these documents in a clean, dry, legible condition, and to make sure they are in good order and available at all times for inspection. The documents are to be labeled "Project Record" in neat large letters and filed concurrently with construction progress. Record documents are not to be used for construction purposes. At the completion of construction they are to be turned over to the owner. In some cases, the contractor also has to develop *as-built* drawings, or as more recently called, record drawings. Frequently, erasable transparencies of the architect's and engineer's drawings are used for this purpose.

The contractor maintains a complete, orderly file, and you should do the same. The contractor usually submits to the owner or consultant at the end of each week copies of his Daily Manpower Log and Daily Progress and Status Notations. This information may be combined on one sheet of paper. Also, he is to submit copies of Construction Photograph monthly with his Request for Progress Payments.

Shop Drawings

There are certain responsibilities associated with shop drawing submittals and review. The contractor's submittal must be timely, he must allow at least 30 days before approved submittals will be needed to allow for possible resubmission and review. The contractor must submit one reproducible transparency and two blue line prints of shop drawings; a sufficient number of copies of product data to satisfy distribution as defined later; one set of samples for selection; and three sets for approval.

Also, submittals must be accompanied by a transmittal letter containing:

1. Project title and number.
2. Contractor's name and address.
3. Date.
4. The number of each shop drawing, product data, and sample submitted.
5. Notification of deviations from contract documents.
6. Other pertinent data.

The actual submittal is to contain:

1. Date and revision date.
2. Project title and number.
3. The names of owner, consultant, contractor, subcontractor, supplier, manufacturer, and detailer.
4. Identification of product or material.
5. Relationship to adjacent structure.
6. Field dimensions.
7. Specification section number.
8. Finishes.
9. Gauges, fastening, reinforcements, welding details.
10. Application standards, such as ASTM or federal specification number.
11. UL or other testing laboratory approval when necessary.
12. Identification of deviation from contract documents.
13. Contractor's stamp, initialed or signed, certifying review of submittal, verification of field measurements, coordination with all trades involved and compliance with contract documents.

The consultant reviews shop drawings but only for conformance with the design concept and with information given in the contract documents. He must expedite his review and usually works to a two-week turnaround period. While this seems to be sufficient time, frequently it requires concerted effort on the part of the consultant to meet the allotted time frame. This is paticularly true where the consultant is an A/E firm and the shop drawing review involves the architect and several engineering disciplines. A good example is a packaged rooftop air-conditioning unit to be installed in a penthouse-type enclosure. The architect must review its dimensions and evaluate it for clearance; structural will check the weight and verify that the supporting members match the dimensions of the unit and will carry the weight; mechanical will review its performance and conformance with the specifications and check for mechanical requirement penetrations; and electrical must verify power requirement.

The specification usually establishes the number of copies that must be submitted and defines the distribution by the contractor as follows:

1. Contractor's file
2. Job-site file
3. Record document file
4. Other prime contractors
5. Subcontractors
6. Supplier
7. Fabricator
8. Owner's file
9. Consultant's file

Operating and Maintenance Manuals

The contractor is to furnish to the owner three sets, individually bound, of clear and complete instructions for proper maintenance of materials, finishes, equipment, machinery, and other items to insure proper care and reasonable life expectancy. These sets are to be printed or typed on $8\frac{1}{2}$ inch × 11-inch sheets, adequately bound in sturdy binders.

The data must include recommendations of specific products and frequency of maintenance in cleaning and lubricating, and replacement items such as filters, product source locations, and servicing agencies. For machinery, maintenance manuals must be provided and include complete parts lists clearly showing the source(s) of genuine replacement parts.

From this it is obvious that construction administration is quite involved, requires complete technical knowledge of all systems, equipment, and materials, and requires experience with the legal aspects of construction.

As was stated at the beginning of this chapter, it is definitely to the advantage of the owner to retain the consultant for construction administration. On larger projects, under additional services, the owner should contract for a full-time on-site representative of the consultant. The fee for these services could be insignificant in comparison to the legal liability of just one false move.

Additionally, the consultant, with his current experience in construction administration could provide valuable assistance in keeping the project on schedule and within budget.

SUMMARY

The best approach to construction administration is to follow a basic step-by-step procedure. This could be helpful even in cases where the consultant performs construction administration, as part of his basic services, or as additional services where full-time on-site administration is provided. In that case you will know what to expect and how it will be administered. In other cases, probably on smaller projects, you will be performing construction administration.

1. Actively participate with the consultant during the various phases of design progress.

2. Review progress drawings and specifications as they are issued for review.

3. Conduct in-depth study of drawings and specifications prior to the start of construction.

4. Participate in the preparation and issue of change orders and be aware of their impact on construction time and costs.

5. Review, to the best of your ability, contractor's cost proposals.

6. Maintain and monitor construction costs on a monthly basis in order to project final total cost. Some corporate executives will not tolerate cost overruns. In order to accommodate changes due to field condition, unforseen problems and delays, and possible owner's changes, a cushion of 5 to 10 percent over the contractor's bid is usually good practice.

7. Make a concerted effort to eliminate, or at least seriously limit owner's changes frequently imposed by changing user needs or executive whims.

8. Contract with the consultant for all basic and additional services deemed necessary to adequately staff the construction administration consistent with the requirements of the project.

9. Mechanical Systems and Equipment

This chapter will address those mechanical systems and equipment that provide for and control man's environment in a building. It will deal with heating, ventilating, and air-conditioning systems, including central chiller equipment, air and water distribution systems, and terminal devices within the spaces. HVAC systems either provide heat to, or extract heat from a space, control humidity, and insure adequate or required ventilation. Depending upon the space(s) and the size of the building, this can be accomplished by any of the following:

- Window air conditioner
- Through-the-wall unit
- Rooftop packaged unit
- Central chiller plant

The central chiller plant system is an extensive array of piping and ductwork, and any number of air handlers and terminal devices whose performance may be controlled by complex microprocessor automation. The selection and application of these systems for effective and efficient operation requires the services of experienced and knowledgeable professionals.

Equipment manufacturers are continually striving to improve their products. This has been going on since scientific air conditioning was introduced in 1902 by Willis H. Carrier and it will continue as long as there is a need for air conditioning. While many improvements in HVAC are the result of on-going research and development, sometimes they are in response to unexpected situations. For example, prior to the energy crisis of 1973, the constant volume, single duct system with terminal reheat was used extensively. Since then, the variable air volume distribution system is being specified almost exclusively. Today, the use of constant volume systems is limited to those situations where more effecient methods cannot be utilized effectively to meet system objectives.

While there have been significant improvements in refrigeration equipment and distribution systems, the basic refrigeration principle has remained the same.

REFRIGERATION CYCLE

The method by which heat is removed from interior spaces is the result of several heat transfer processes in accordance with certain laws of thermodynamics. The evaporation of a liquid and the condensation of a gas or vapor are known as the process in which heat is absorbed and rejected.

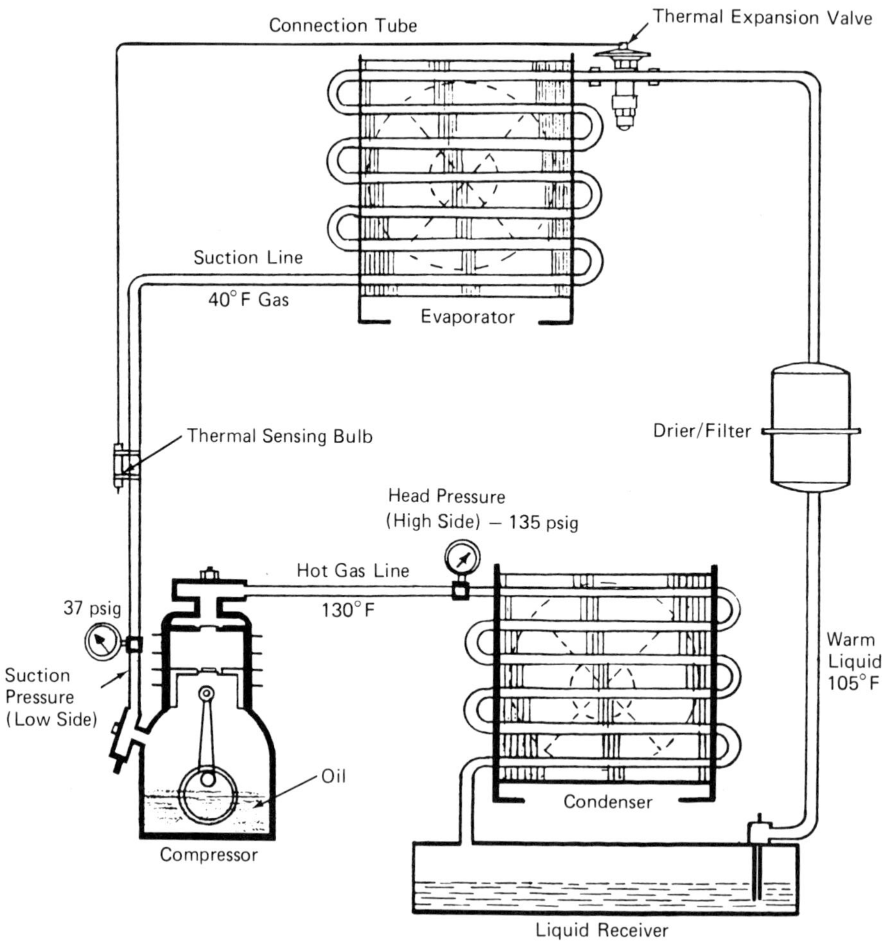

Figure 9-1. Typical reciprocating refrigeration system.

Figure 9–1 shows a typical reciprocating refrigeration system consisting of a compressor, condenser, expansion valve, evaporator, fans, drier/filter, and control. Safety control devices are not shown as they are not necessary to illustrate the refrigeration principle. The actual pressures and temperatures may vary somewhat depending upon the system and the refrigerant used, but it does illustrate how the system functions, and the relationship between the various pressures and temperatures. The evaporator is in the space to be cooled or, if remote, the air to be cooled is ducted through the evaporator, and the condenser is usually located outdoors. The evaporator absorbs the heat from the air and this heat is rejected through the condenser. The maximum exchange of heat that can theoretically take place between the refrigerant liquid as it approaches the expansion valve and the vapor in the evaporator is limited by the difference in their specific heats. The quantity of heat that each pound of refrigerant liquid absorbs while flowing through the evaporator is known as the *refrigerating effect*.

The compressor is the heart of the refrigerating system; it continuously circulates the refrigerant through the system, and its piston displacement determines the capacity of the refrigerating system as a whole.

This type of system is used in packaged units such as: window units; through-the-wall units; rooftop unitary or split systems.

Figure 9-2 illustrates a typical centrifugal package chiller with a compressor, shell

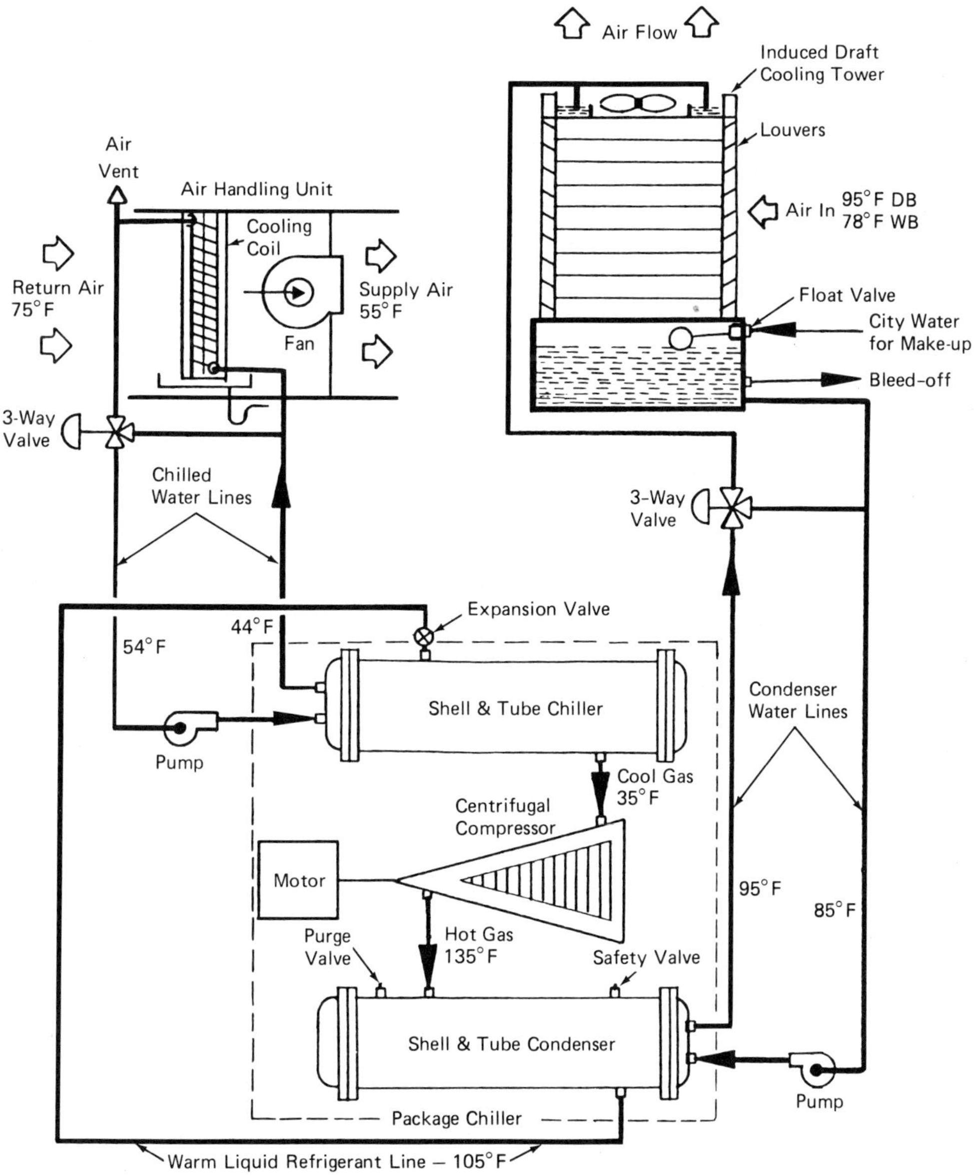

Figure 9-2. Typical central centrifugal chiller system.

and tube heat exchanger, air handling unit, and cooling tower. Again, the temperatures are typical and not necessarily those found in all actual installations.

With this system, the air-cooled condenser is replaced by a shell and tube condenser and the heat is rejected through a remote cooling tower. The evaporator (direct expansion, DX) coil is replaced by a shell and tube chiller (evaporator) and an air-handling unit. The refrigeration principle remains the same.

These systems are used in larger installations or wherever it is more economical to pipe chilled water than duct cooled air. Also, because of space limitation, it would be impractical to duct air in large ducts as compared to piping water. Further, this permits the use of any number of air-handling units for zone control or differing temperature requirements.

While this illustration uses 3-way valves in both the chilled water and condenser water lines, not all systems are specified with 3-way valves or locate them as shown. In some installations the condenser water is bypassed into the cooling tower basin instead of the return line. In some instances, an automatic 2-way control valve is installed in the discharge side of the air handling unit coil. As long as certain design criteria are satisfied in the chiller and condenser, the valving flow control may be as best suits those criteria.

REFRIGERATION SYSTEM

There are three basic types of compressors used in refrigeration systems, and these cover a broad range of capacities from the small 1/4-ton unit window air conditioner up to several thousand tons for large centrifugal machines. Good engineering practice dictates the use of several smaller machines instead of one large unit. Smaller machines can be placed on-line, one after another, as required to satisfy the constantly varying cooling load. The air-conditioning load does vary, by the hour of the day and the day of the year. The maximum total cooling load in the northern hemisphere usually occurs in the late afternoon during the middle of August depending upon the type of building (roof area as compared to wall area), amount and type of glass and exposure, and the type of occupancy. For example, office buildings have a different peak load period from restaurants and these both differ from theaters. Solar heat gain may peak at 9 A.M. for a certain exposure or at 6 P.M. for another, whereas for a roof, it may be at 2 P.M.

With several smaller machines, the failure of any one may have no significant effect on cooling capacity except on a hot summer afternoon.

Generally, reciprocating compressors are available in capacities up to 200 tons, centrifugal machines from 100 tons up to about 8000 tons, and the helical rotary (screw) compressors range from 125–800 tons.

Another type is the absorption liquid chiller which is heat-operated and is primarily limited to installations where surplus heat is normally available.

The water chiller (evaporator) in a central plant system is a shell and tube exchanger and performs essentially the same function as the DX coil does in small packaged units, but with one exception: here the heat transfer medium is water. The evaporator produces chilled water that is then pumped to air handlers, fan coil units, etc., located at some remote point from the central plant, and in or adjacent to the space to be cooled. While Figure 9–2 shows a single duct constant volume, fixed temperature system with terminal reheat for temperature control, there are other means for controlling space temperature.

The heat absorbed by the remote cooling coil is transferred through the shell and tube condenser to the cooling tower where it is rejected to the atmosphere. Exactly the same principle as the simple window air conditioner except the system is much larger and far more complex.

In order to determine the performance of an air-conditioning system, appropriate instrumentation is required. This could consist of thermometers judiciously placed to determine entering and leaving temperatures at the various points of heat transfer, or in some cases Btu meters are used. This instrumentation is used by the manufacturer of the equipment and/or the contractor to verify that performance meets design criteria and can be used by building maintenance personnel to monitor performance during the life of the equipment.

There are two basic formulas that can be used to calculate heat transfer in water circuits (Specific heat of water = 1):

$$H = Q \times TD \times 500$$

where:
H = Heat transferred in Btu/hr
Q = Water quantity in gpm
TD = Temperature difference, °F

Since 12,000 Btu = 1 ton of refrigeration, the above formula can be converted to:

$$\text{Tons} = Q \times TD/24$$

In order to utilize these formulas, it is necessary to know the water flow in gpm. This requires the installation of flow meters in appropriate locations.

The quantity of heat rejected by the condenser exceeds the quantity of heat extracted by the evaporator by the quantity of heat added during the compression process. Generally about 240 Btu per minute (14,400 Btu per hour) is rejected to the condenser per ton of refrigeration capacity.

In the past, chillers usually operated on a 10°F temperature drop, 52–42°F. This has changed, in the interest of economy, to the point where 58–44°F is not unusual. Before being overly concerned about the various temperatures encountered, find out exactly what design conditions are for the specific chiller and system. In the condenser, 85–95°F water temperature is common.

The typical specifications included in this and the following chapter are condensed and intended only as a guide listing important features and requirements. All references to specific manufacturers or suppliers have been purposely deleted. All specifications (except for state and federally funded projects) for such equipment as chillers, cooling towers, air-handling units, pumps, valves, etc., usually include one or more manufacturers to indicate level of quality required. As stated in chapter 8, specifications may be written for:

Figure 9-3. Factory assembled package chiller mounted on a steel base. The steel base is isolated from concrete equipment pad by vibration isolators to prevent transmission of both vibration through the structure and sound through the system. Condenser water piping with butterfly valves at the rear of the machine and power controller is set on housekeeping pad in the foreground.

- One manufacturer with no substitution
- Two or more manufacturers with no substitution
- One or more manufacturers or approved equal
- No manufacturer, any product meeting specifications being acceptable

An actual refrigeration specification on a recently designed $30 million office building consisted of 25 articles covering 15 pages. The inclusion of complete mechanical and electrical specifications is beyond the intended scope of this book.

TYPICAL SPECIFICATION
Refrigeration

This section describes the design, operating features, and installation of the refrigeration system, auxiliary equipment, starter, motor, and controls required for the complete central refrigeration plant to provide chilled water for air-conditioning systems. Two of the five machines shall be equipped with heat recovery bundles.

Supply and install in the location shown a centrifugal refrigeration system which shall be the standard design of the manufacturer. Each system shall be complete and include a centrifugal compressor, hermetic motor, cooler, condenser, refrigerant flow control device, purge unit, and/or transfer unit and refrigerant storage receiver (not with R-11 refrigerant), and control center with related system mounted controls. The system shall be factor assembled, piped, wired, leak tested, evacuated, and charged with refrigerant.

Systems that operate at subatmospheric pressure (R-11, R-113) shall be provided with a complete purge unit, providing positive means for collection, return of refrigerant, and removal of noncondensables. Systems that operate above 15 psig (R-12, R-22, R-114, R-500) shall be provided with a complete compressor operated transfer unit and storage receiver to permit transfer and isolation of the full refrigerant charge independent of system compressor operation. Where condenser shell is a separate vessel of ample size for storage of the entire refrigerant charge, and is furnished with isolation valves, a separate storage receiver and compressor for transfer of refrigerant is not required.

It is the intent of this section that insofar as possible manufacturer's standard equipment, of a type which has been in successful operation for at least five years, shall be used. It is not the intent to specify completely all details of design, control, and construction of equipment, but the type, size, and materials of this equipment shall be subject to approval of required shop drawing submittal. Accordingly, this specification must be considered an outline, and all accessories, piping connections, thermostats, gauges, and other equipment and apparatus which are necessary for the complete operation of the refrigeration machine shall be furnished the same as though specified herein.

(Complete technical information is usually inserted at this point.)

The manufacturer's engineer shall perform the initial start-up of the system including proper coordination with the contractor on start-up of the cooling tower and associated condenser and chilled water pumps.

The contractor shall provide a competent operating force to operate the equipment for a period of three 8-hour days until it is accepted by the owner.

The chiller manufacturer's engineer shall instruct the owner in the proper care and operation of the system for a period of five days.

Each contractor shall furnish to the owner clear and complete instructions for proper maintenance, operation, and spare parts list of equipment and systems furnished under this contract. Data shall be triplicate, typed on $8\frac{1}{2} \times 11$-inch sheets, and adequately bound in separate sturdy binders. Data shall contain the name, address, and phone number of the closest representative who can be contacted in case of need. Information about manufacturer's service contracts shall be made available to the owner.

COOLING TOWERS

There are two types of condensers used with air-conditioning systems. The air-cooled condenser is used with smaller packaged units, through-the-wall units, and the common window air conditioner. In this case, air is the heat transfer medium. The larger, central plant refrigeration systems use water-cooled condensers. Here, water is the heat transfer medium.

In addition to the heat added in the evaporator (heat absorbed from the space cooled), the heat added to the refrigerant during the compression process, must be carried away. In the past, and in some isolated cases even today with smaller installations, water is used on a once-through basis. This method is now seldom used. In many localities the cost of water is too high to permit wasting it after it has been used. In other places, the water available for condensing purposes is so warm that relatively large quantities are required in order to keep the condenser pressure down to a

Figure 9-4. Multi-cell cooling tower with condenser water distribution piping above hot water basin. Flow control valves along left railing are adjusted to provide uniform distribution over hot water basin. Heavy gauge galvanized wire screen is installed on each fan stack. Fan drive motor is in the foreground.

moderate level. In either case, the cost of condenser water for a refrigerating plant may be so high that it is desirable to use the same water over and over again, cooling the water after it leaves the condenser.

The most common method of accomplishing this is with the familiar cooling tower. There are two broad categories of cooling towers:

- The natural draft in which the air movement through the tower depends upon natural wind movements.
- The forced (or induced) draft tower through which air is forced (or drawn) by fans

While each has its own unique advantages and disadvantages, the forced (or induced) draft tower is most frequently used for building air-conditioning systems. Water is pumped through the condenser, usually a shell-and-tube heat exchanger, where the water passes through the tubes and the hot gas from the compressor enters at the top of the condenser and the liquid refrigerant exits from the bottom. The water is pumped under pressure through the condenser up to the hot water basin at the top of the cooling tower. The water is then evenly distributed over the tower fill area where the water comes in contact with the forced or induced airstream. The cooled

Figure 9-5. Louvers direct inlet air-flow across interior fill material on 3-cell cooling tower. Spring vibration isolators installed between dunnage and structural steel. Ladder with safety cage provides access to the top of the tower.

water collects in the tower sump and is returned to the pump for continuously circulating operation.

The water, as it passes through the tower, is cooled by evaporation. The temperature of the water will approach, but never reach, the wet bulb temperature of the air. A small portion of the water, about 1 percent, is evaporated. Also, some water is captured (entrained) by the airstream, but the quantity carried outside the tower is reduced by eliminators. Properly placed eliminators will restrict the loss to less than $\frac{1}{2}$ of 1 percent.

Water lost to evaporation and entrainment is replaced by make-up. This is accomplished by a float in the basin that may either control the float valve directly or remotely.

Special precautions are necessary when cooling towers are to be operated in subfreezing weather. Condenser water must be diverted directly to the tower basin and heaters must be installed to maintain at least 40°F. Steam or electric basin heaters are most commonly used. All exposed piping must be protected (heat traced) or drained when the tower is operated intermittently during cold weather.

In most cases, water treatment is necessary to prevent scaling, corrosion, or biological fouling of the condenser and circulating system because evaporation of the water causes a buildup of dissolved solids. Sometimes overflow or blowdown is used to control this buildup.

Figure 9-6. A 2-circuit heat tracing control panel automatically maintains condenser water at 40° F temperature in subfreezing weather. Indicator lights monitor power to circuit and thermostat position.

TYPICAL SPECIFICATION
Cooling Tower

This section describes the performance, erection, and installation of cooling towers.

The side casing shall be of galvanized sheet steel. Tower framing shall be hot dipped galvanized steel. All bolts and hardware shall be hot dipped galvanized steel. The tower shall be constructed to withstand a wind pressure of 30 lbs/sq ft equivalent to a 100 mile/hour wind. All material shall be noncombustible.

Factory installed heaters shall be provided to maintain 40°F in the sump when outside ambient is 0°F at a 15 mph wind.

Tower shall be constructed and erected so that it is water tight when in operation. Maximum total make-up requirements shall be limited to a full load evaporation loss of 1 percent plus drift losses of not more than .2 of 1 percent.

(Complete technical information on fan, motor, piping, noise levels, water treatment, and safety are usually inserted at this point)

AIR HANDLING UNITS

Air handling units, as addressed here, may be either factory-assembled or site-fabricated, and may contain: heating and cooling coils, supply and return fans, motors, filters, humidifier, mixing plenum, freezestats and firestats, outside, return and relief air dampers, fire and smoke dampers, and associated indicating devices and control provisions.

Air handling, as such, is also performed in window air conditioners, through-the-wall units, fan coil units, and packaged rooftop HVAC system. In each case, the function of the air handling portion is the same, providing conditioned air to the space in a quantity sufficient to maintain design conditions. Window units usually merely cool and filter the air. Fan coil units may heat or cool and filter the air, and provide outside ventilation air. Large systems usually provide completely conditioned air, including heating, cooling, humidification, dehumidification, filtration, and ventilation in either constant or variable air quantities. The function of the air handler is to provide this conditioned air to the space through appropriate distribution system. Air handling units are designed to match the air distribution system and these are categorized according to the means by which heating/cooling is accomplished in the conditioned space:

- Single-duct with terminal reheat, constant volume
- Double-duct, constant volume
- Single-duct, variable volume
- Multizone, constant volume

While a constant volume fan can satisfy all of the above requirements, including single-duct variable volume, by appropriate bypass dampers, variable volume fans are usually used in this application.

On smaller air-handling systems only supply fans are used and the mixture of return and outside is controlled by automatic dampers. On larger systems, both supply and return fans are used.

The conventional systems such as double duct, multizone, and terminal reheat, all using constant volume air delivery, consume a large amount of the total HVAC construction cost and are generally unsatisfactory, in large system application, because of their inefficient use of energy. In order to meet the challenge of reducing first cost while making better use of energy, an old system was reborn—the variable volume system. It offers the following advantages:

- Savings in initial costs due to reduced equipment and duct sizes.
- Savings in operating costs, by reducing horsepower, particularly at part load conditions.
- Savings in space due to smaller equipment sizes.
- It is essentially self-balancing.

Several items to look for in mechanical room and air-handling unit design drawings are:

- Adequate and appropriately placed lighting for maintenance and servicing of equipment.
- Floor drains for use in the event coils must be drained and for drainage from condensate pans.
- Adequately sized access doors into air handlers for maintenance and service.
- Fan-bearing lubrication fittings must be extended from inaccessible bearings for servicing.

- Lighting within the housing where required.
- Isolation valves on supply and return lines of all coils.
- Unions or flanged connections at all coils for ease of removal.
- Freezestat and firestat inclusion.
- Thermometer locations for ease of reading from floor location at supply and return side of all coils.
- Mechanical or automatic air vents in the water lines at coils.
- Vibration isolation for proper installation, motion not inhibited.
- Proper mix of return air and outside air to eliminate stratification.

There are several formulas that can be used to calculate heat transfer in air circuits:

Sensible Heat, Btu/hr $= Q \times 1.08 \times TD$
Total Heat, Btu/hr $= Q \times 4.5 \times HD$
Latent Heat, Btu/hr $= Q \times 0.68 \times WD$
Sensible Heat Ratio = Sensible Heat/Total Heat
where:
Q = Air Quantity, cfm (cubic feet/minute)
TD = Temperature difference, °F
HD = Enthalpy of air difference, Btu/lb
WD = Weight of moisture difference, grains (or pounds) of water/lb of dry air

Other useful air side formulas:

Air changes/hr $= Q \times 60/V$
Velocity of air, feet/min $= Q/A$

where:

V = Volume of space in cubic feet
A = Area of duct in square feet (To determine face velocity of a coil or filter, use the area of the coil or filter)

Fan brake hp $= Q \times TP/6350 \times FME$

where:

TP = Total pressure, in inches of water
FME = Fan mechanical efficiency

Selection and Application of Fans

Although fans are manufactured in various types, fans used in air handling units fall into two general categories, axial and centrifugal, depending on the direction of airflow through the impeller.

Figure 9-7. Single inlet, backward-curved blade centrifugal fan. (Courtesy of Buffalo Forge Company)

Centrifugal fans are the workhorse of the industry because of their versatility and low first cost. The wheel or rotor revolves in the housing, pulling air in around one or both ends of the wheel axis and discharging it through the housing outlet. Blades are either radial, forward-curved, backward-inclined, or air foil. Variable volume is accomplished either by varying the speed of the fan or by controlling the air by variable air inlet vanes.

Noise emited by centrifugal fans can be easily controlled by simple attenuation devices inserted into the duct system.

The total-pressure rise developed by a fan is proportional to the tangential component of the absolute exit velocity. The absolute exit velocity is the sum of the velocity relative to the wheel plus the tip speed of the wheel. In the radial-tip wheel the absolute exit velocity of the air always exceeds the tip speed of the wheel. With the backward curved wheel it is less, and with the forward curved wheel it may be considerably more. It is, therefore, possible for a fan wheel to discharge air at a velocity higher than its tip speed. The power demand for the radial-tip wheel rises steadily with increasing flow; for the forward wheel the increase is more rapid, whereas for the backward wheel the increase passes through a maximum then decreases. For a fixed quantity of air flow the forward curved wheel fan will operate at a slower speed than the backward curved wheel fan to produce the same total pressure. The backward curved fan will generally use less horsepower than the other two.

Axial-flow fans are versatile, sometimes less costly than centrifugal fans, and are

Figure 9-8. Fixed-blade vaneaxial fan. (Courtesy of Buffalo Forge Company)

finding increased application because of the development of controllable-pitch blades.

There are three types of axial flow fans (propeller, tubeaxial, and vaneaxial fans). They are distinguished by the duty for which they are designed and not by blade shapes.

Propeller Fans. They are most often used for ventilating commercial and industrial space, and are mounted in the wall with little or no duct work. They could be either direct drive or belt drive and are for applications usually with less than one inch static pressure. The power required is lowest at maximum flow and highest at zero flow. For the same initial cost, propeller fans will deliver larger volumes with less power required.

Tubeaxial Fans. Basically, these are just heavy duty propeller fans. The clearance between the blade and the tube should be the minimum practical value in order to limit leakage. Since the air is discharged from the fan blade with a spiral motion, it is poor practice to discharge through long ducts unless special precaution is taken. This spiral motion produces higher duct losses than does straight air flow as from a centrifugal fan. Because of their higher noise level these fans are usually limited in application to where noise is tolerable.

Vaneaxial Fans. These are tubeaxial fans with discharge vanes that straighten out the spiral motion. These discharge vanes actually transform the kinetic energy produced by the propeller in the form of a whirl into more useful pressure energy. To make this transformation efficiently, the air discharged from the propeller must be guided through a gradual turn until the tangential component is eliminated.

It sometimes becomes necessary to effect changes in fan performance. For instance, due to changes in conditions, a larger or lesser quantity of air must be moved by the same fan. The solution to this problem does not necessarily mean the purchase of a new fan to satisfy the new conditions.

The performance of the same fan for different speeds and for different densities, and the performance of similar fans operating at the same tip speed and for the same densities can be predicted by the fan laws.

1. The same fan handling the same gas (air, in all cases discussed) at different speeds:

Quantity (cfm) varies directly with fan speed.

$$Q_2 = (N_2/N_1) \times Q_1$$

Head (static pressure developed) varies directly as the speed squared.

$$H_2 = (N_2/N_1)^2 \times H_1$$

Horsepower varies as the speed cubed.

$$HP_2 = (N_2/N_1)^3 \times HP_1$$

Example: A fan delivers 10,000 cfm with a fan speed of 400 rpm against a head of 0.6 inch of water using 2 hp. Calculate Q, H, and HP for 500 rpm.

$$\begin{aligned} Q &= (500/400) \times 10{,}000 = 12{,}500 \text{ CFM} \\ H &= (500/400)^2 \times 0.6 = 0.94 \text{ INCHES} \\ HP &= (500/400)^3 \times 2 = 3.91 \text{ HP} \end{aligned}$$

Calculations are to slide rule accuracy, which is adequate for practical purposes because, since the motor would not be designed to match the requirement, the next larger size would be selected.

2. Similar fans of different diameters and the same tip speed.

Q varies directly as the diameter squared
HP varies directly as the diameter squared
H remains the same

3. Similar fans of different diameters and the same shaft speed.

Q varies directly as the diameter cubed
HP varies directly as the diameter to the fifth power
H varies directly as the diameter squared

From all the above it is obvious that no one fan can satisfy all requirements, nor is there one fan for any special need. Fan selection is usually a matter of choosing the best size and type from those available for the particular application.

Fan selection is a procedure which begins with specifications of requirements and ends with the evaluation of alternative possibilities. Of the many fans which may be capable of satisfying a particular capacity and pressure requirement, the best selection is the one that does the job most economically. In addition to the requirements imposed by the system, first cost, operating costs, and maintenance costs must be considered.

TYPICAL SPECIFICATION
Air Handling Units

Plenums and Casings

Plenums and casings for site-fabricated air handling units shall include all closures for, and air passages between, the air intakes, filters, heating and cooling coils, and fans. The floors of strucural platforms under casings shall serve as the bottoms, and the sides and tops shall be constructed of sheet metal.

Access doors shall be provided in each section of the casing and shall be of airtight construction. Door frames shall be welded in place, and each door shall be neoprene gasketed, have a minimum of two brass hinges, and be fastened with a minimum of two brass tension fasteners operable from inside and outside the casing.

Fans

The cfm capacity shown in the fan schedules is the minimum the fan shall deliver at the resistance shown.

The resistances, as shown in the schedules, are those which will be required by the ductwork and apparatus and do not include the internal and intake fan losses and outlet velocity heads.

Fans shall not operate at speeds in excess of 80 percent of their true critical speed.

Fan manufacturers shall check WR^2 value of the fan to be certain that the speed-torque curve of the motor specified will accelerate the fan wheel to design speed without overload. Two speed fans will start on low speed and that WR^2 value will be applicable.

Each fan shall be dynamically balanced at the manufacturer's plant before shipment.

Fans shall operate quietly and without pulsation. Manufacturer shall submit sound power level for each fan in accordance with AMCA Standard 301.

Whenever a fan is equipped with inlet vane and outlet damper, submit fan curves showing the relation between static pressure, cfm and bhp at 25, 50 and 75 percent open.

Where shown, provide adjustable vanes mounted on the fan inlet. The number of vanes shall be one less than the number of blades in the fan wheel, and vanes shall be operated from a center mechanism. Each vane shall be supported at the ends by machined bronze bearings. Where vanes are to be operated manually, a means of securely clamping in the desired position shall be provided. When automatic operation is indicated, the proper arm for easy connection of damper motor linkage shall be furnished.

Chilled and Hot Water Coils

Water coils shall be of the number of rows and arrangement, and meet the performance shown on the schedule.

Water coils shall be fin-and-tube type constructed of seamless copper tubes and copper or aluminum fins mechanically bonded or soldered, or helically wound to tubes.

Tubes shall be correctly circuited for proper water velocity without excessive pressure drop and shall be drainable where required or indicated on drawings. Each coil shall be tested at the factory under water at not less than 250 psi air pressure and shall be suitable for 200 psi working pressure. Coils shall be mounted for counterflow service. (Drainable coils shall be installed in the air handling units with a pitch of not less than $\frac{1}{8}$ inch per foot of tube length toward the drain end.) Coils shall conform to the provisions of ARI Standard 410. Water coil capacities, pressure drops, and selection procedures shall be certified in accordance with ARI Standard 410.

Each cooling coil section shall be provided with a stainless steel drip pan not less than 18 gauge with connection to floor drain.

Each cooling coil having an air velocity over 400 feet per minute through the net face area shall be equipped with moisture eliminators unless the coil manufacturer guarantees that no moisture will be carried beyond the drip pans under actual conditions of operation. Where eliminators are included, auxiliary drip troughs shall be provided.

V-Belt Drives

Belts shall be matched sets of premium quality, endless cord, impregnated rubber of trapezoidal cross-section. Belts shall be multibelt, except for utility sets less than 1000 cfm, designed for 150 percent of specified motor nameplate rating.

Belt Guards

Each V-belt drive shall have a belt guard constructed of #16 gauge galvanized flat expanded $\frac{1}{2} \times 1$-inch metal tack welded or bolted into a galvanized angle frame securely bolted to base or apparatus. Guard shall completely enclose drives and pulleys and be constructed in accordance with OSHA and all applicable safety codes.

Provide 4-inch diameter holes, centered at motor and fan shaft with a 6-inch diameter cover plate, adequately secured and constructed so that it can be rotated for access to shaft.

Filters

Filters shall be throwaway type complete with viscose coated glass filament media. Media shall not slump or shed. It shall be dimensionally stable and not give off odor.

Prefilter shall be of the automatic renewable media type, of glass filter, nominal thickness of 2 inch. Material shall be precharged with an adhesive and wound on a spool to form a compact roll. Media shall be automatically fed over the face of the filter and rerolled. Each spool shall be provided with a motor and drive assembly. A pressure switch control shall advance the media as dust accumulation dictates. A fan interlock shall insure no media is fed when the system is inoperative.

After filters shall have a deep-pleated filtering medium with an effective area of 48 square feet per 1000 cfm. The percentage of open area shall not be less than 90 percent of the total face area. For filter efficiencies see schedule.

Filter Gauge

The differential pressure gauge for all filters shall be diaphram operated for continuous indication on a 4-inch dial. The gauge shall have two static pressure tips, two 5 foot lengths of $\frac{1}{2}$-inch aluminum tubing, two compression fittings and mounting plates.

Humidifier

Unit shall be a self-contained, completely packaged type consisting of an electric steam generator, dry steam duct mounted humidifier, and all required controls.

AIR DISTRIBUTION SYSTEM

A most important link between the central chiller (heating) plant and the building spaces is the air distribution system. In order to maintain desired conditions, adequate quantities of air must be delivered to these spaces. Since the proper quality of the air is a function of the air handling unit, this section addresses quantity only. The proper quantity must be delivered to the space with minimum noise, without creating drafty conditions, and with uniform temperature distribution throughout.

After the consultant has determined the heat gain/loss for the space, he designs an air distribution system consisting of properly selected duct sizes, with appropriate auxiliary devices for controlling and directing the airstream, to insure adequate air quantities at the terminal devices. The entire air distribution system, when completely installed, will be balanced to satisfy design conditions.

As explained earlier, there are several different types of air distribution systems in common use. Probably the simplest of all is the single zone system. This includes the window air conditioner, the through-the-wall unit and the rooftop unit. The unit may be installed within the space or the air may be ducted from a remotely located unit. The most common application of the single-zone system is in small spaces such as classrooms, small individual shops in shopping malls, in small plant spaces. Sometimes, a number of individual units are installed in a larger area where each unit functions as a single zone system. These units may be installed with or without distribution ductwork.

Control for single zone system can be accomplished by cycling refrigeration, face and bypass dampers, with terminal reheat, or a combination of these.

Another system, used almost exclusively in the past, is the constant volume with terminal reheat. The air is cooled by the air handling unit to a preset value (with the advent of microprocessors, this temperature can now be varied in accordance with load conditions and outside temperature), ducted throughout the building and reheated at various locations near or in the space to be conditioned. Individual thermostats are used to control the reheat coils to maintain desired temperatures. Reheaters may be steam, hot water, or electric.

The constant volume, terminal reheat system is energy intensive. The fan system must move maximum design quantity of air even during partial load conditions and the temperature of the air leaving the air handling unit must be cooled to the lowest temperature required for any one zone. Air handling units for these systems frequently use what is known as an economizer cycle, using outside air when its heat content is less than that of the return air, for free cooling.

While variable air volume systems are not suitable for every application, they are being specified more frequently with each passing year. It is an accepted fact that VAV systems consume less energy than conventional systems such as double duct, multizone, and constant volume terminal reheat.

VAV systems vary the air quantity to a space at a constant temperature to balance the heat gains or losses to achieve and maintain the desired room temperature. Operating cost savings are obtained by reducing supply (and return) air quantities, reducing horsepower required, and thereby reducing power and energy consumption. Further economies are obtained from reduced equipment and duct sizes.

Terminal Devices

The true VAV system uses a variable volume fan system and variable volume terminal device. Some systems use constant volume fans with VAV terminals. Some VAV terminals utilize reheat coils in those applications where added heat is required to compensate for excessive heat loss.

VAV terminals, sometimes called "boxes," are available in several different types:

- Variable volume, single duct primary air
- Variable volume, single duct primary air with reheat coil
- Variable volume bypass type, single duct constant volume
- Variable volume, dual duct mixing box
- Variable volume induction box

The most common type in use today is the variable volume box, with or without reheat, using constant temperature primary supply air in a single duct. A space thermostat directly controls the supply air quantity. This system is used with variable volume fans whose output is controlled by a pressure sensor located at a remote point from the fan to maintain constant pressure in the ducts. As space temperature changes, the space thermostat will modulate the control of the VAV box changing the air quantity as needed. This change in air quantity will be sensed as a pressure change by the pressure sensor in the duct and will in turn modulate the variable volume fan output.

These VAV boxes are frequently factory set to the maximum design CFM and some minimum quantity, usually depending upon ventilation requirements. Balancing these systems is as simple as balancing the constant volume system.

The variable volume, bypass-type box performs exactly as the name implies. It is supplied by a constant volume air handling unit with air at a temperature low enough to satisfy the maximum room/zone cooling load. This temperature may be held constant or reset as stated earlier. Any air that is not required to maintain space temperature is bypassed to the return air system which may consist of a ceiling plenum (codes permitting), or a return air duct. While this provides a simple means for space temperature control, there is no energy savings as the fan requires constant horsepower to provide constant volume.

The variable volume, induction box operates in the reverse of the above except that the output of the box is essentially constant. The induction box is supplied with conditioned primary air at a temperature that will handle design space cooling load. The volume of primary air is varied to suit load conditions. At maximum cooling, the box will provide 100 percent primary air. When the space thermostat senses a reduction in load, the volume of primary air is decreased and secondary air is induced. This secondary air could be room air or plenum air (codes permitting). The temperature of blended air is always such that it will satisfy load conditions. While the box delivers essentially constant volume air to the space, it must be connected to a variable volume fan system. The main advantage of this box is that it provides constant volume of air to the space to insure design air circulation within the space. While most diffusers are designed to perform well with variable volume systems, they are usually selected on basis of volume of air handled and the throw of the diffuser. Constant volume through them assures a most consistent mixing of the air within the space. The added cost of the induction box over the true VAV box may be too great to offset its advantage.

Diffusers, Grilles, and Registers

In addition to supplying the proper quality and quantity of air to the space, it is also necessary to distribute the air within the space as uniformly as practical with minimum draft sensation by the occupant and to provide as complete an air mix as possible with minimum noise. The ideal situation is created by laminar flow where the air moves uniformly across the space either up, down, or across. Unfortunately, this method is not used for office space air conditioning.

Modern air diffuser technology has been developed to the point where ideal conditions are almost attainable. Space occupants usually complain more frequently about drafts, stratification, and noise as a result of improper diffuser selection than about the temperature of the space. It is simple enough to dump air into a space to control its temperature, but to properly diffuse the air to satisfy the occupant is another matter.

The ideal condition is most nearly satisfied by a properly designed 4-way blow diffuser placed in the center of the ceiling of a square room whose dimensions match the throw distance of a diffuser. The type of diffuser and its location within the space determines its ultimate performance.

Supply diffusers, grilles, and registers are available in numerous types and sizes:

- Circular diffusers
- Square and rectangular diffusers
    - 1-way blow
    - 2-way blow
    - 3-way blow
    - 4-way blow
- Linear diffusers
- Grilles and registers
    - Square
    - Rectangular
    - Linear
- Luminaire diffusers

Since these devices are predesigned, neither the consultant nor the owner can do much more than select from those available. Further, since there are many designs, types, and sizes to choose from, the proper selection and judicious application become the determining factors of ultimate performance.

The primary function of these diffusers is to introduce the air into the space in such a manner as to provide thorough mixing with the space air so as to create no noticeable sensations. The air entering the space could be as low as 55°F in the summer and as high as 120°F in the winter. A ceiling diffuser will direct the airflow along the ceiling (protrusions such as soffits, beams, surface mounted luminaires, etc., may prematurely divert the airstream into the occupied area) for a distance equal to the "throw" of the diffuser. At this point, the airstream velocity will be reduced to the "terminal velocity" in feet per minute that exists at the end of the throw of the diffuser. Diffuser selection based upon proper throw and terminal velocity will best suit the conditions required for the space.

Sheet Metal Duct

Last, but certainly not least in importance, is the sheet metal ductwork. The function of this ductwork is to convey the conditioned air from the air handling unit through the space diffuser in the most economical manner. Sheet metal ductwork, unlike fans, diffusers, etc., which are predesigned and fabricated and require only selection and application, is designed for a specific application, shop fabricated, and field installed. Therefore, design considerations are appropriate at this point.

A duct system imposes resistance to the airflow which must be overcome by the expenditure of mechanical energy which is supplied by a fan. The resistance of a system is a function of the volume of air flowing through the system. This volume depends upon the fan selected and its performance characteristics. The fan will deliver a specific volume of air against a specific duct system pressure. Fan/duct pressures are small and are normally expressed in inches of water. At any cross-section in a duct the total pressure P_t is the sum of the static pressure P_s and the velocity pressure P_v. Actually, only the total and static pressure can be measured, and the velocity pressure is found from their difference.

In HVAC systems, standard air having a specific volume of 13.13 cu ft/lb, i.e., a specific weight of 0.075 lb/cu ft is used to simplify the formula for velocity pressure:

$$P_v = (V/4005)^2$$

where: V = mean velocity of flow, feet per minute and:

$$V = Q/A$$

where: Q = air flow rate, cubic feet per minute
A = cross-sectional area of duct, square feet
then: $P_t = P_s + P_v$

Whenever air flows through a duct, some pressure is lost due to internal friction of the air and the movement of the air along the surface of the duct. The loss of pressure due to friction is called the friction head of the duct. A fan moves air through the ducts by providing sufficient pressure to overcome the friction losses in the duct work.

Fan performance must match system requirements. The only possible operating points are those where the system characteristics intersect the fan characteristics. At that point the pressure developed by the fan exactly matches the system resistance and the flow through the system equals the capacity.

Typical fan performance curves show fan capacity in cubic feet per minute (cfm) plotted along the abscissa and static (or total) pressure, fan speed in revolutions per minute (rpm), horsepower, efficiency, and sound level along the ordinate. Once the cfm of a fan/system is determined all the other characteristics of the fan can be easily picked off the curves for that particular fan.

The first cost of a duct depends upon its size. The smaller the duct, the lower will be its first cost. However, the power required to deliver the air through the smaller duct rises because of the increase in pressure due to the increased friction loss.

The best duct design is the one that produces the desired results most economically within acceptable noise limits of the application. Duct design is a trade-off of owning costs against operating cost. Increasing duct size increases installed cost but decreases operating cost and, of course, the converse is also true.

There are four common design methods for sizing ducts:

Constant-Velocity. This is used in simple systems without branches and in those exhaust systems in which the material transported might settle if the velocity were reduced.

Velocity-Reduction. This may be used in designing supply or exhaust systems having numerous branches. For designing supply systems this method consists of selecting an appropriate velocity for the first section of main, and progressively decreasing the velocity at each take-off. Since in this method the branch velocity is arbitrarily chosen as a certain fraction of that in the main, it is necessary to check whether the available pressure is equal to or greater than the loss for the design flow. Dampers may be used for balancing when the loss is less than the available pressure. Velocity reduction in the direction of flow on supply systems is also achieved automatically.

Equal-Friction. This method is employed in designing a variety of supply and exhaust systems. It is most effective for symmetrical systems or for systems where the lengths of all the various runs are approximately equal. For such systems balancing is automatically achieved by designing for equal friction per foot length.

Static-Regain. This may be used for designing supply or exhaust systems having numerous branches, each connected to a single relatively long main. It utilizes the increase in static pressure (which accompanies any velocity reduction) to provide equal static pressure at each take-off or function. If the length of main between branches is either very long or very short, this method may not be practical.

The *equal-friction method* and the *velocity-reduction method* are the two methods most commonly used in designing duct systems.

Round ducts require less metal to carry a given quantity of air than rectangular ducts; however, quite often rectangular ducts are used because of space considerations. Duct design is usually done for the round duct, and in cases where rectangular

Table 9-1. Design Velocities for Conventional and High Velocity Systems.

DUCT ELEMENT	CONVENTIONAL PUBLIC BUILDINGS		HIGH-VELOCITY COMMERCIAL BUILDINGS	
	NORMAL	MAX	NORMAL	MAX
Main Ducts	1000	1600	2500	6000
Branch Ducts	800	1300	2000	4500
Fan Outlets	1500	2200	2500	5000
Suction Inlet	800	1100	1700	3300
Outside Air Intakes	500	900	600	1000
Filters	300	350	350	350
Heating Coils	500	600	600	700
Air Washers	500	—	500	—
Dehumidifiers	500	—	500	—

Courtesy of Buffalo Forge Company

ducts are required, the round is converted to the equivalent rectangular shape. The friction loss in the equivalent rectangular duct is the same as for the round duct for the same quantity of air.

Example: Design a duct system for 20,000 cfm, 200 feet in length with four equal outlets spaced at 50 feet and the friction loss not to exceed 0.2 inches of water.

Friction loss is equal to 0.1 inches of water per 100 feet. Use Figure 9–9 and read vertically from 0.1 inch to 20,000 cfm and find duct diameter 43 inches with 2,000 fpm velocity. After the first outlet, the quantity is reduced to 15,000 cfm and the duct diameter (from chart) is reduced to 38 inches with 1,900 fpm. After the second outlet the quantity is reduced to 10,000 cfm, duct diameter is 35 inches with a velocity of 1,700 fpm. After the last section of duct before the fourth outlet, the quantity is 5,000 cfm, duct diameter is $25\frac{1}{2}$ inches and the velocity is 1,450 fpm. The equal-friction provides a simple and rapid solution to duct design problems. This round duct can now be converted to equivalent rectangular duct, should that be necessary, by using Table 9–2.

TYPICAL SPECIFICATION
Sheet Metal Ductwork

Sheet Metal

This section describes the construction of ductwork, manufactured of galvanized sheet steel unless noted otherwise, and the methods to be used in the intallation. This contractor shall install all automatic dampers supplied by the control contractor.

All sheet metal work and accessories shall conform with latest edition of construction standards as applicable and as published by Sheet Metal and Air Conditioning Contractors National Association, Inc. (SMACNA)

Samples, shop, erection, and fabrication drawings for all work of this section shall be submitted for approval before fabrication is started.

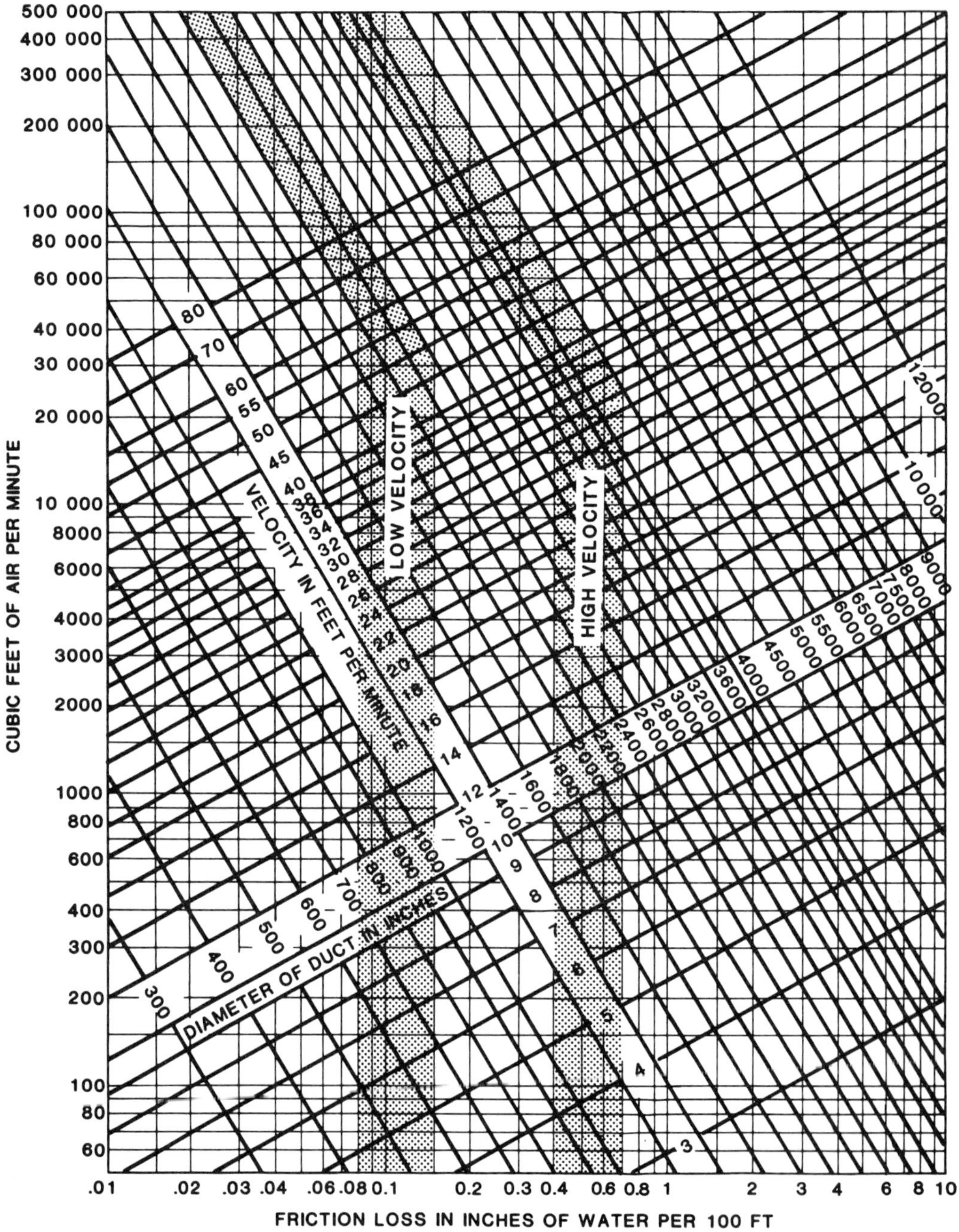

Figure 9-9. Shaded area indicates suggested velocity and friction rate design limits. To use, read vertically from Friction Loss to the intersection of horizontal projection from CFM and find velocity in FPM and round duct diameter in inches. (Reprinted by permission from ASHRAE, *Handbook of Fundamentals*, 1981)

Table 9-2 Circular Equivalents of Rectangular Ducts for Equal Friction and Capacity

Dimensions in Inches

Side Rectangular Duct	4.0	4.5	5.0	5.5	6.0	6.5	7.0	7.5	8.0	9.0	10.0	11.0	12.0	13.0	14.0	15.0	16.0
3.0	3.8	4.0	4.2	4.4	4.6	4.7	4.9	5.1	5.2	5.5	5.7	6.0	6.2	6.4	6.6	6.8	7.0
3.5	4.1	4.3	4.6	4.8	5.0	5.2	5.3	5.5	5.7	6.0	6.3	6.5	6.8	7.0	7.2	7.5	7.7
4.0	4.4	4.6	4.9	5.1	5.3	5.5	5.7	5.9	6.1	6.4	6.7	7.0	7.3	7.6	7.8	8.0	8.3
4.5	4.6	4.9	5.2	5.4	5.7	5.9	6.1	6.3	6.5	6.9	7.2	7.5	7.8	8.1	8.4	8.6	8.8
5.0	4.9	5.2	5.5	5.7	6.0	6.2	6.4	6.7	6.9	7.3	7.6	8.0	8.3	8.6	8.9	9.1	9.4
5.5	5.1	5.4	5.7	6.0	6.3	6.5	6.8	7.0	7.2	7.6	8.0	8.4	8.7	9.0	9.3	9.6	9.9

Side Rectangular Duct	6	7	8	9	10	11	12	13	14	15	16	17	18	19	20	22	24	26	28	30	Side Rectangular Duct
6	6.6																				6
7	7.1	7.7																			7
8	7.6	8.2	8.7																		8
9	8.0	8.7	9.3	9.8																	9
10	8.4	9.1	9.8	10.4	10.9																10
11	8.8	9.5	10.2	10.9	11.5	12.0															11
12	9.1	9.9	10.7	11.3	12.0	12.6	13.1														12
13	9.5	10.3	11.1	11.8	12.4	13.1	13.7	14.2													13
14	9.8	10.7	11.5	12.2	12.9	13.5	14.2	14.7	15.3												14
15	10.1	11.0	11.8	12.6	13.3	14.0	14.6	15.3	15.8	16.4											15
16	10.4	11.3	12.2	13.0	13.7	14.4	15.1	15.7	16.4	16.9	17.5										16
17	10.7	11.6	12.5	13.4	14.1	14.9	15.6	16.2	16.8	17.4	18.0	18.6									17
18	11.0	11.9	12.9	13.7	14.5	15.3	16.0	16.7	17.3	17.9	18.5	19.1	19.7								18
19	11.2	12.2	13.2	14.1	14.9	15.7	16.4	17.1	17.8	18.4	19.0	19.6	20.2	20.8							19
20	11.5	12.5	13.5	14.4	15.2	16.0	16.8	17.5	18.2	18.9	19.5	20.1	20.7	21.3	21.9						20
22	12.0	13.0	14.1	15.0	15.9	16.8	17.6	18.3	19.1	19.8	20.4	21.1	21.7	22.3	22.9	24.0					22
24	12.4	13.5	14.6	15.6	16.5	17.4	18.3	19.1	19.9	20.6	21.3	22.0	22.7	23.3	23.9	25.1	26.2				24
26	12.8	14.0	15.1	16.2	17.1	18.1	19.0	19.8	20.6	21.4	22.1	22.9	23.5	24.2	24.9	26.1	27.3	28.4			26
28	13.2	14.5	15.6	16.7	17.7	18.7	19.6	20.5	21.3	22.1	22.9	23.7	24.4	25.1	25.8	27.1	28.3	29.5	30.6		28
30	13.6	14.9	16.1	17.2	18.3	19.3	20.2	21.1	22.0	22.9	23.7	24.4	25.2	25.9	26.6	28.0	29.3	30.5	31.7	32.8	30
32	14.0	15.3	16.5	17.7	18.8	19.8	20.8	21.8	22.7	23.5	24.4	25.2	26.0	26.7	27.5	28.9	30.2	31.5	32.7	33.9	32
34	14.4	15.7	17.0	18.2	19.3	20.4	21.4	22.4	23.3	24.2	25.1	25.9	26.7	27.5	28.3	29.7	31.0	32.4	33.7	34.9	34
36	14.7	16.1	17.4	18.6	19.8	20.9	21.9	22.9	23.9	24.8	25.7	26.6	27.4	28.2	29.0	30.5	32.0	33.3	34.6	35.9	36
38	15.0	16.5	17.8	19.0	20.2	21.4	22.4	23.5	24.5	25.4	26.4	27.2	28.1	28.9	29.8	31.3	32.8	34.2	35.6	36.8	38
40	15.3	16.8	18.2	19.5	20.7	21.8	22.9	24.0	25.0	26.0	27.0	27.9	28.8	29.6	30.5	32.1	33.6	35.1	36.4	37.8	40
42	15.6	17.1	18.5	19.9	21.1	22.3	23.4	24.5	25.6	26.6	27.6	28.5	29.4	30.3	31.2	32.8	34.4	35.9	37.3	38.7	42
44	15.9	17.5	18.9	20.3	21.5	22.7	23.9	25.0	26.1	27.1	28.1	29.1	30.0	30.9	31.8	33.5	35.1	36.7	38.1	39.5	44
46	16.2	17.8	19.3	20.6	21.9	23.2	24.4	25.5	26.6	27.7	28.7	29.7	30.6	31.6	32.5	34.2	35.9	37.4	38.9	40.4	46
48	16.5	18.1	19.6	21.0	22.3	23.6	24.8	26.0	27.1	28.2	29.2	30.2	31.2	32.2	33.1	34.9	36.6	38.2	39.7	41.2	48
50	16.8	18.4	19.9	21.4	22.7	24.0	25.2	26.4	27.6	28.7	29.8	30.8	31.8	32.8	33.7	35.5	37.2	38.9	40.5	42.0	50
52	17.1	18.7	20.2	21.7	23.1	24.4	25.7	26.9	28.0	29.2	30.3	31.3	32.3	33.3	34.3	36.2	37.9	39.6	41.2	42.8	52
54	17.3	19.0	20.6	22.0	23.5	24.8	26.1	27.3	28.5	29.7	30.8	31.8	32.9	33.9	34.9	36.8	38.6	40.3	41.9	43.5	54
56	17.6	19.3	20.9	22.4	23.8	25.2	26.5	27.7	28.9	30.1	31.2	32.3	33.4	34.4	35.4	37.4	39.2	41.0	42.7	44.3	56
58	17.8	19.5	21.2	22.7	24.2	25.5	26.9	28.2	29.4	30.6	31.7	32.8	33.9	35.0	36.0	38.0	39.8	41.6	43.3	45.0	58
60	18.1	19.8	21.5	23.0	24.5	25.9	27.3	28.6	29.8	31.0	32.2	33.3	34.4	35.5	36.5	38.5	40.4	42.3	44.0	45.7	60
62		20.1	21.7	23.3	24.8	26.3	27.6	28.9	30.2	31.5	32.6	33.8	34.9	36.0	37.1	39.1	41.0	42.9	44.7	46.4	62
64		20.3	22.0	23.6	25.1	26.6	28.0	29.3	30.6	31.9	33.1	34.3	35.4	36.5	37.6	39.6	41.6	43.5	45.3	47.1	64
66		20.6	22.3	23.9	25.5	26.9	28.4	29.7	31.0	32.3	33.5	34.7	35.9	37.0	38.1	40.2	42.2	44.1	46.0	47.7	66
68		20.8	22.6	24.2	25.8	27.3	28.7	30.1	31.4	32.7	33.9	35.2	36.3	37.5	38.6	40.7	42.8	44.7	46.6	48.4	68
70		21.0	22.8	24.5	26.1	27.6	29.1	30.4	31.8	33.1	34.4	35.6	36.8	37.9	39.1	41.2	43.3	45.3	47.2	49.0	70
72			23.1	24.8	26.4	27.9	29.4	30.8	32.2	33.5	34.8	36.0	37.2	38.4	39.5	41.7	43.8	45.8	47.8	49.6	72
74			23.3	25.1	26.7	28.2	29.7	31.2	32.5	33.9	35.2	36.4	37.7	38.8	40.0	42.2	44.4	46.4	48.4	50.3	74
76			23.6	25.3	27.0	28.5	30.0	31.5	32.9	34.3	35.6	36.8	38.1	39.3	40.5	42.7	44.9	47.0	48.9	50.9	76
78			23.8	25.6	27.3	28.8	30.4	31.8	33.3	34.6	36.0	37.2	38.5	39.7	40.9	43.2	45.4	47.5	49.5	51.4	78
80			24.1	25.8	27.5	29.1	30.7	32.2	33.6	35.0	36.3	37.6	38.9	40.2	41.4	43.7	45.9	48.0	50.1	52.0	80
82				26.1	27.8	29.4	31.0	32.5	34.0	35.4	36.1	38.0	39.3	40.6	41.8	44.1	46.4	48.5	50.6	52.6	82
84				26.4	28.1	29.7	31.3	32.8	34.3	35.7	37.1	38.4	39.7	41.0	42.2	44.6	46.9	49.0	51.1	53.2	84
86				26.6	28.3	30.0	31.6	33.1	34.6	36.1	37.4	38.8	40.1	41.4	42.6	45.0	47.3	49.6	51.6	53.7	86
88				26.9	28.6	30.3	31.9	33.4	34.9	36.4	37.8	39.2	40.5	41.8	43.1	45.5	47.8	50.0	52.2	54.3	88
90				27.1	28.9	30.6	32.2	33.8	35.3	36.7	38.2	39.5	40.9	42.2	43.5	45.9	48.3	50.5	52.7	54.8	90
92					29.1	30.8	32.5	34.1	35.6	37.1	38.5	39.9	41.3	42.6	43.9	46.4	48.7	51.0	53.2	55.3	92
96					29.6	31.4	33.0	34.7	36.2	37.7	39.2	40.6	42.0	43.3	44.7	47.2	49.6	52.0	54.2	56.4	96

(Reprinted by permission from ASHRAE *Handbook of Fundamentals*, 1981)

Table 9-2 Circular Equivalents of Rectangular Ducts for Equal Friction and Capacity (*Concluded*)

Dimensions in Inches

Side Rectangular Duct	32	34	36	38	40	42	44	46	48	50	52	56	60	64	68	72	76	80	84	88	Side Rectangular Duct
32	35.0																				32
34	36.1	37.2																			34
36	37.1	38.2	39.4																		36
38	38.1	39.3	40.4	41.5																	38
40	39.0	40.3	41.5	42.6	43.7																40
42	40.0	41.2	42.5	43.7	44.8	45.9															42
44	40.9	42.2	43.5	44.7	45.8	47.0	48.1														44
46	41.8	43.1	44.4	45.7	46.9	48.0	49.2	50.3													46
48	42.6	44.0	45.3	46.6	47.9	49.1	50.2	51.4	52.5												48
50	43.6	44.9	46.2	47.5	48.8	50.0	51.2	52.4	53.6	54.7											50
52	44.3	45.7	47.1	48.4	49.7	51.0	52.2	53.4	54.6	55.7	56.8										52
54	45.1	46.5	48.0	49.3	50.7	52.0	53.2	54.4	55.6	56.8	57.9										54
56	45.8	47.3	48.8	50.2	51.6	52.9	54.2	55.4	56.6	57.8	59.0	61.2									56
58	46.6	48.1	49.6	51.0	52.4	53.8	55.1	56.4	57.6	58.8	60.0	62.3									58
60	47.3	48.9	50.4	51.9	53.3	54.7	60.0	57.3	58.6	59.8	61.0	63.4	65.6								60
62	48.0	49.6	51.2	52.7	54.1	55.5	56.9	58.2	59.5	60.8	62.0	64.4	66.7								62
64	48.7	50.4	51.9	53.5	54.9	56.4	57.8	59.1	60.4	61.7	63.0	65.4	67.7	70.0							64
66	49.4	51.1	52.7	54.2	55.7	57.2	58.6	60.0	61.3	62.6	63.9	66.4	68.8	71.0							66
68	50.1	51.8	53.4	55.0	56.5	58.0	59.4	60.8	62.2	63.6	64.9	67.4	69.8	72.1	74.3						68
70	50.8	52.5	54.1	55.7	57.3	58.8	60.3	61.7	63.1	64.4	65.8	68.3	70.8	73.2	75.4						70
72	51.4	53.2	54.8	56.5	58.0	59.6	61.1	62.5	63.9	65.3	66.7	69.3	71.8	74.2	76.5	78.7					72
74	52.1	53.8	55.5	57.2	58.8	60.3	61.9	63.3	64.8	66.2	67.5	70.2	72.7	75.2	77.5	79.8					74
76	52.7	54.5	56.2	57.9	59.5	61.1	62.6	64.1	65.6	67.0	68.4	71.1	73.7	76.2	78.6	80.9	83.1				76
78	53.3	55.1	56.9	58.6	60.2	61.8	63.4	64.9	66.4	67.9	69.3	72.0	74.6	77.1	79.6	81.9	84.2				78
80	53.9	55.8	57.5	59.3	60.9	62.6	64.1	65.7	67.2	68.7	70.1	72.9	75.4	78.1	80.6	82.9	85.2	87.5			80
82	54.5	56.4	58.2	59.9	61.6	63.3	64.9	66.5	68.0	69.5	70.9	73.7	76.4	79.0	81.5	84.0	86.3	88.5			82
84	55.1	57.0	58.8	60.6	62.3	64.0	65.6	67.2	68.7	70.3	71.7	74.6	77.3	80.0	82.5	85.0	87.3	89.6	91.8		84
86	55.7	57.6	59.4	61.2	63.0	64.7	66.3	67.9	69.5	71.0	72.5	75.4	78.2	80.9	83.5	85.9	88.3	90.7	92.9		86
88	56.3	58.2	60.1	61.9	63.6	65.4	67.0	68.7	70.2	71.8	73.3	76.3	79.1	81.8	84.4	86.9	89.3	91.7	94.0	96.2	88
90	56.8	58.8	60.7	62.5	64.3	66.0	67.7	69.4	71.0	72.6	74.1	77.1	79.9	82.7	85.3	87.9	90.3	92.7	95.0	97.3	90
92	57.4	59.3	61.3	63.1	64.9	66.7	68.4	70.1	71.7	73.3	74.9	77.9	80.8	83.5	86.2	88.8	91.3	93.7	96.1	98.4	92
94	57.9	59.9	61.9	63.7	65.6	67.4	69.1	70.8	72.4	74.0	75.6	78.7	81.6	84.4	87.1	89.7	92.3	94.7	97.1	99.4	94
96	58.4	60.5	62.4	64.3	66.2	68.0	69.7	71.5	73.1	74.8	76.3	79.4	82.4	85.3	88.0	90.7	93.2	95.7	98.1	100.5	96

Equation for Circular Equivalent of a Rectangular Duct:

$$D_e = 1.30\,[(ab)^{0.625}/(a + b)^{0.250}]$$

where

a = length of one side of rectangular duct, mm.
b = length of adjacent side of rectangular duct, mm.
D_e = circular equivalent of rectangular duct for equal friction and capacity, mm.

(Reprinted by permission from ASHRAE *Handbook of Fundamentals*, 1981)

All ductwork shall be free from pulsation, noise, chatter, vibration, or objectionable noise. Any of these defects that occur after system is in operation shall be corrected.

The sound absorbtion coefficient for low velocity duct lining shall be in accordance with the following table:

Octave Band Center Frequency in H_z	125	250	500	1000	2000	4000
Noise Reduction Coeff. (NRC)	0.30	0.52	0.64	0.87	0.87	0.80

The lining shall be glass fiber and odorless, minimum 3 lb density per cubic foot, coated on the air stream side with neoprene or vinyl. Shall be 1 inch thick and have a "k" factor of 0.23 at 75°F mean temperature.

Lining adhesive and coating shall have a fire and smoke hazard rating not exceeding a flame spread of 25, fuel contributed 50, and smoke developed 50, and shall not support mold growth.

To prevent lining from flaking off, all joints and seams shall be filled with adhesive, and entering and leaving edges shall be protected by metal "Zee" edge attached to the ductwork.

In order to prevent sound passing through the area remaining between duct and opening in floors and walls, mineral wool shall be packed to completely fill the opening. Fireproof rope shall be used when fire rated walls or floors are penetrated. Galvanized sheet metal safe-off shall be used to close the gap between the structure and the duct.

Dishwasher Exhaust Duct

Duct shall be constructed of a minimum #20 gauge, ASTM A167, #304 stainless steel, with all seams and joints completely welded. Pitch ductwork back to dishwasher and provide $\frac{3}{4}$-inch copper tubing to drain low points to open waste.

Flexible Ducts to Air Terminals

The flexible air ducts between the supply ducts and the air terminals shall be constructed of corrugated ductile aluminum that can be bent and rebent by hand and is self-supporting. Maximum length shall be limited to 5 feet or as indicated by governing codes. Insulation shall be fitted over the outside and have an exterior vinyl cover.

Flexible Connections

On the inlet and outlet connection of each fan or unit and at building expansion joints, there shall be installed a flexible connection of not less than 3-inch wide of flexible fire retardant, closely woven glass fabric double coated with neoprene.

Low Velocity, Low Pressure Duct

This section describes the installation of low velocity, low pressure ductwork. (2-inch maximum static pressure and 2000 ft/min maximum velocity)

All sheet metal gauges for ductwork, reinforcing angles, standing seams and plenum chambers shall conform to SMACNA *Low Velocity Duct Construction,* latest edition.

Provide turning vanes where shown on drawings and in all square elbows.

Air extractors shall consist of a series of radius vanes pivoted to operating bars and bars to be fastened to an operating bracket with means for locking into any fixed position.

Volume dampers shall be constructed of not less than #22 gauge. Where installed in ducts up to 12 inches deep, dampers shall be single blade. In ducts deeper than 12 inches, dampers shall be multiblade, each blade being not wider than 9 inches.

Ducts shall be adequately supported using trapeze or strap hangers. Perforated band iron will not be permitted. Ducts shall not be supported from furring, hung ceilings, or from another duct or pipe.

High Velocity, Medium Pressure Ductwork

This section describes the intallation of high velocity, medium pressure supply air duct system (2–6 inch static pressure and 2000 ft/min velocity or greater) materials and methods to be used in their installation.

All ductwork shall conform to NFPA No. 90A, and SMACNA standards except where modified herein.

Air leakage shall not exceed 1 percent of volume when tested at 125 percent of the operating static pressure.

Acoustically line both duct and fittings, where indicated, with fiberglass with a thermal conductivity of 0.27 Btu/hr/sq ft/°F at 75°F mean temperature. Lining shall be covered with zinc-coated perforated sheet steel with an open area of 25 percent minimum.

Where high velocity ductwork is not continuously welded, joints shall be made airtight with sealants.

Supply ductwork which is not acoustically lined shall be insulated (wrapped) with 2-inch thick fiberglass blanket, 1 lb density with factory applied vapor barrier facing. The thermal conductivity shall be 0.29 Btu/hr/sq ft/°F at 75°F mean temperature. Circumferential joints shall butt, longitudinal joints shall overlap 2 inches. Joints shall be covered with 3-inch wide vapor barrier foil reinforced tape.

TYPICAL SPECIFICATION

Diffusers, grilles, registers, and louvers

Fixed exterior louvers and screens shall be provided by the general contractor. This contractor shall connect ductwork, where required, to make a watertight connection.

Grilles and registers shall be of one manufacturer, of the size, design, and construction to handle the amount of air as shown.

Ceiling diffusers shall be provided with an adjustable grid designed to distribute the air evenly over the entire face of the outlet and have an adjustable volume damper in the collar.

TESTING AND BALANCING

Testing and balancing is the final phase of an HVAC system installation. In the past, when systems were simpler, this was frequently done by the mechanical contractor. Today, this work is more frequently performed by a testing and balancing contractor under a separate contract directly with the owner.

Testing and balancing is necessary to match the performance of the installed system to design conditions. Most problems in HVAC systems are a result of inadequate or improper air and water distribution. The engineer that designs a system bases the anticipated results on the fact that certain quantities of air and water will flow through various heat exchangers and terminal devices. Unless these conditions are met, the desired results cannot be obtained.

Under the separate contract method, which is discussed here, both the installing contractor (mechanical contractor and his appropriate subcontractors and vendors) and the t & b contractor have certain specific responsibilities. The only way these contractors will know what is expected is by being told in the specifications.

Under General Requirements, a section is inserted to indicate that "Owner shall employ and pay for the services of an independent Testing and Balancing Contractor." Then, under Division 15, a similar statement is included and the responsibilities of the mechanical contractor are defined. Finally, usually Division 17 is the specification for testing and balancing. While it is not part of the contract and the mechanical contractor will not bid this phase of work, he knows what is to be bid separately. At some later date, usually shortly after the general contract is let, the owner will seek bids from contractors to perform testing and balancing. This division is freestanding and also includes the responsibilities of the mechanical contractor exactly as described in those sections. In this manner, every involved party is fully knowledgeable about what is expected of everyone else.

TYPICAL SPECIFICATION

Testing and Balancing

This contractor shall provide all labor, engineering, and test equipment not furnished by the mechanical contractor but that is required to adjust and balance the entire HVAC system. All work shall be under the direct supervision of a registered professional engineer.

This contractor shall cooperate with the mechanical contractor to the fullest extent possible to assure a complete well-balanced system.

Air system testing requirements:

1. Test and adjust all fan speeds to deliver required cfm.
2. Make pitot tube traverse of main supply ducts to verify design cfm.
3. Test and adjust for design outside air.
4. Test and adjust for design recirculated air.
5. Test and adjust each diffuser, grille, and register to within 5 percent of design requirements

Hydronic system testing requirements:

1. Prior to testing and balancing, obtain a representative water sample, have a chemical analysis performed, and include the results in the final report.
2. Check all flow meters for proper installation, calibration and accuracy.
3. Measure and adjust circulating water pump flow capacity to proper quantity.
4. Adjust flow of chilled water through chillers.
5. Adjust flow of hot water through boilers.
6. Balance chilled and hot water system flows.
7. Adjust flow of condenser water.

Testing and balancing data to be recorded on standard forms and four copies submitted to owner at the completion of work. All data forms shall include all pertinent nameplate data and relevant associated data for comparison, for example:

Fans:

- Manufacturer
- Model or serial number
- Type and number of blades
- Cfm, rated, and measured
- Rpm, Rated, and measured
- Pressure, measured
- Pulley size, type, and manufacturer
- Belt size and quantity

For motors record rated and full load running current. For airflow record specified and actual cfm.

This contractor shall conduct sound level readings, using approved instruments in areas specified ____

____.

Upon completion of testing and balancing, this contractor shall replace all adjustable pitch sheaves with fixed pitch sheaves of the proper size. These fixed pitch sheaves shall be furnished by the mechanical contractor and he shall turn over to the owner all adjustable pitch sheaves.

Responsibility of the mechanical contractor:

(This section has been included in the specifications of the mechanical contractor and repeated here for the testing and balancing contractor for the sake of continuity and coordination between contracts.)

1. At least two months in advance, submit a schedule listing the date of completion of his work as it is ready for testing and balancing.
2. Two weeks prior to the start of balancing activities, attend a coordinating meeting between the owner/engineer, and balancing contractor, and continue on a scheduled weekly basis until completion of the work. This meeting will be conducted under the guidance of the general contractor.
3. Certify that, prior to the start of t & b work, all systems are complete, operational, and ready for t & b, and that a rough balancing has been completed.
4. Place all automatic, hand, and balancing valves in the open position.
5. Remove and clean all strainers.
6. Verify that expansion tanks are operating properly.
7. Install adjustable pitch sheaves on all fans and after completion of balancing furnish fixed pitch sheaves to the t & b contractor sized as specified.
8. Furnish to the t & b contractor portable airflow measuring devices, portable water flow meters, portable water flow indicators, mercury thermometers, pressure guages, and other devices as required.
9. All technical data furnished with the equipment provided by the mechanical contractor, and appropriate drawings shall be submitted to the t & b contractor.
10. Provide all necessary labor to operate mechanical equipment during testing and balancing.
11. Make all modifications to rectify discrepencies reported by the t & b contractor as indicating noncompliance with the contract documents.
12. Be responsible for the protection of all mechanical devices, equipment, and systems provided under his contract during the testing and balancing period.

VALVES

In HVAC and plumbing systems, valves are used to not only control the flow of water, but the rate, volume, pressure, and direction; that is, they can turn on and off, regulate, modulate, or isolate. Valves are either manual or automatic, quarter-turn or multiturn, with or without actuators, or rising or nonrising stem types. The basic material from which valves are manufactured are bronze, iron, and steel, in which there are the carbon, forged, and stainless materials, or a valve could be made of an exotic metal alloy. Also, valves are available with a number of different means for connection to pipe, such as: soldered, screwed, flanged, and welded. This selection depends upon the piping system used. For a valve to be a long-lasting mechanism, it must combine the proper body material with the right seating and packing materials to resist wear and corrosion throughout the life of the valve.

Because of the variety of valves, various methods are used to type, group, and classify them. For the purposes of this discussion, valves will be grouped according to valve handle operation, and further subdivided as to means of flow control.

1. Multiturn valves
   a. Gate
   b. Globe
   c. Diaphram
   d. Needle
2. Quarter-turn valves
   a. Plug
   b. Ball
   c. Butterfly
3. Self-actuated
   a. Relief
   b. Check
      i. Swing
      ii. Lift
      iii. Backflow preventer
   c. Pressure reducing

(Back-flow preventer is a special device, usually contains two check valves with a unique center chamber, designed to isolate the treated [or contaminated] water of HVAC systems from the potable water system. These are specifically designed for the application and usually must be accepted by the authority having jurisdiction.)

The multiturn valve further divides into rising and nonrising stem types with either inside screw or outside screw. The rising stem has one distinct advantage over its counterpart in that the position of the valve can be determined by the position of the stem. Further, in fire protection (sprinkler) systems, valves that control the water supply are required to have tamper switches that monitor the position of these valves. The rising stem provides the necessary means for the actuation of the tamper switch.

Gate valves, as the name implies, usually have a wedge-shaped disc that moves perpendicular to the direction of fluid flow. The wedge shape, tapered in the direction of closure, slides into a matched seat in the valve body to form a tight, metal-to-metal seal, and is able to withstand elevated temperatures and pressures. Gate valves permit almost full flow, minimum pressure drop, and minimum turbulence. They are used where operation is infrequent and should not be used for throttling. Gate valves are available with flat discs and solid, split, or hollow wedges. They are available with nonrising stem and inside screw, rising stem and inside screw, and rising stem and outside screw.

Globe valves are the most versatile of all valves and use a plug moving against the direction of fluid flow to mate with a seat in the valve body. They can be used to throttle or regulate fluid flow as well as stop it. In water systems, all globe valves are in-

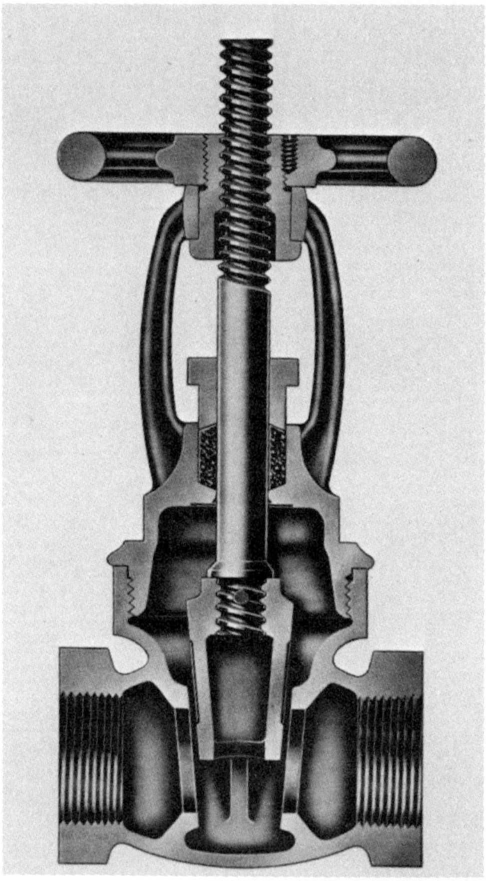

Figure 9-10. Gate valve with rising stem and outside screw and yoke (OS&Y) design. Rising stem gives direct indication of gate position inside the valve. (Courtesy of Jenkins Bros.)

stalled with the inlet pressure *beneath* the seat; however, when operating conditions are very severe (for example, high temperature steam) it is sometimes best to install with pressure *above* the seat. The pressure drop is much more pronounced than in a gate valve. There are many choices of disc types to consider if throttling service is a real consideration, since one disc may significantly outlast another under varying conditions. Conventional discs or plug-type discs (which give optimum throttling performance) can be made of bronze, iron, stainless steel, nickel alloy, or other materials, where operation is frequent and absolute tightness is essential, composition discs are available. Renewable discs or regrind–renew types with renewable seat rings are also available.

Diaphram valves use a flexible element that is mechanically deflected to control the flow of the fluid. This is ideal for handling difficult fluids because the operating mechanism is isolated from the fluid. Diaphram valves are not used in HVAC systems.

Needle valves are a variation of the globe valves and use a tapered needle inserted in

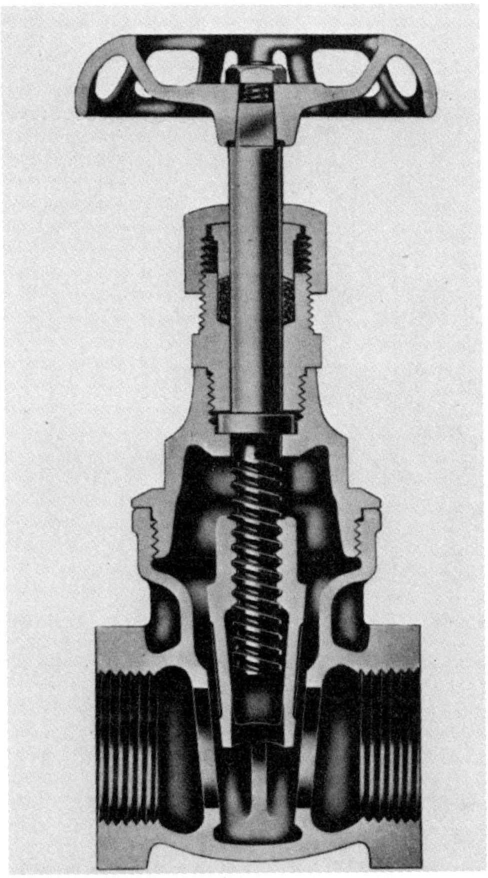

Figure 9-11. Gate valve with non-rising stem and inside screw requires minimum head room for operation. Gate position cannot be determined by observation of the handwheel. (Courtesy of Jenkins Bros.)

an orifice to control flow. They are used to throttle fluid flow but do not find much application in HVAC systems.

Plug valves are probably the oldest type of control valve in use for on/off, non-throttling, flow control. It has a tapered plug with an opening through it that is rotated in the valve body to control the fluid flow. By design, port size is varied from about 50 to 100 percent, with 70 percent being most common. The tapered plug provides tight shutoff because of the seating characteristics of the mating tapered sealing surfaces.

Ball valves are similar to plug valves except they use a ball with a hole through one axis to connect the inlet and outlet ports in the body. The 40-year-old ball valve is basically an adaptation of the several-thousand-year-old plug valve. The ball valve has little pressure drop, since it has a straight-through port which gives unobstructed flow. Throttling is possible but can cause rapid wear, and service should be primarily on-off. The modern ball valve improves upon the older plug type by having much lower turning torque, no metal-to-metal contact since it rotates between two resilient

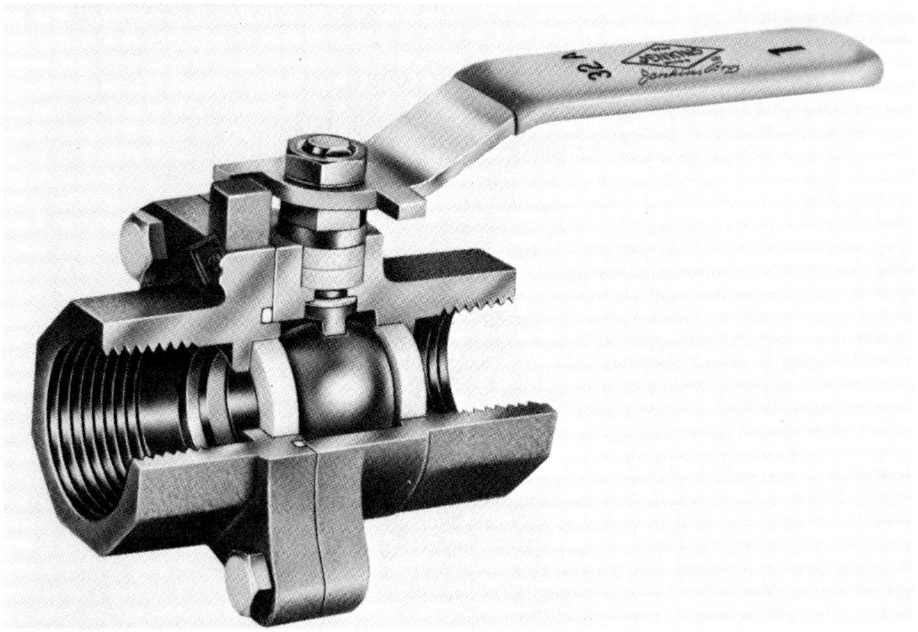

Figure 9-12. Bronze ball valve with TFE seats, packing and gaskets, and stainless steel ball and stem. (Courtesy of Jenkins Bros.)

seats with concave seating surfaces. Most common seating surfaces are Teflon (for saturated steam) and Buna-N (for general services).

Fluid flow is straight through in the open position (either full, part, or reduced port opening) and a 90° rotation from the open position completely blocks the flow of fluid. Like plug and gate valves, ball valves are not generally recommended for throttling services.

Ball valves are excellent for on-off service because they are light in weight and simple in construction, are low in cost, offer minimum resistance to flow and provide a tight shut-off. Because they can be opened so quickly, ball valves may be responsible for water hammer or surge pressures in systems where they are installed. These valves find frequent application in HVAC and plumbing systems.

Butterfly valves control flow by using a circular disc rotating at right angles to the direction of fluid flow. When the plane of the disc is parallel to direction of fluid flow, the valve is fully open. When the disc (handle) is rotated 90° the valve is closed. As with all quarter-turn valves, care must be exercised in opening and closing butterfly valves to avoid water hammer. These valves have low pressure drop, are light in weight, require minimum space, and are used for on-off and throttling service.

A typical 6-inch wafer butterfly valve measures just over 2 inches flange-to-flange and weighs about 30 lbs, whereas a similar sized flanged gate valve is over 10 inches long and weighs 175 lbs. Elastometric seats provide bubble-tight seal for butterfly valves.

As the disc is moved from the fully open position, it offers progressively more resistance to flow hence requires more force near the fully closed position. For throt-

Figure 9-13. Lug-type butterfly valves with manual operators, left to right; without operator; locking lever operator; gear operator with position indicator 2″-4″; gear operator with position indicator 5″-14″, babbitt sprocket rim. (Courtesy of Jenkins Bros.)

tling purposes, the handle is provided with locking means to hold the disc in the selected position. Although butterfly valves can be used for throttling purposes, its primary use is for on-off service. They are used quite commonly in HVAC application.

Self-actuated valves are designed to operate under abnormal conditions or to maintain normal conditions. Check valves are installed in pipelines to prevent fluid flow in a direction opposite to the normal intended flow. Relief valves are designed to permit fluid flow in cases of excessive pressure buildup. This fluid is relieved to atmosphere

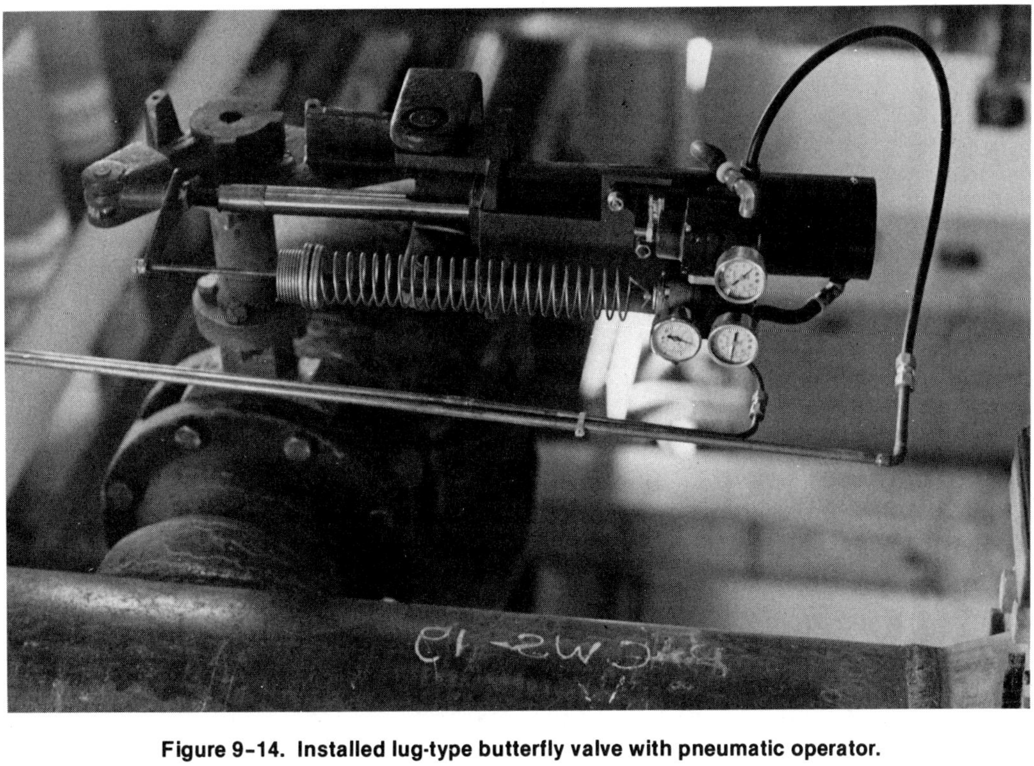

Figure 9-14. Installed lug-type butterfly valve with pneumatic operator.

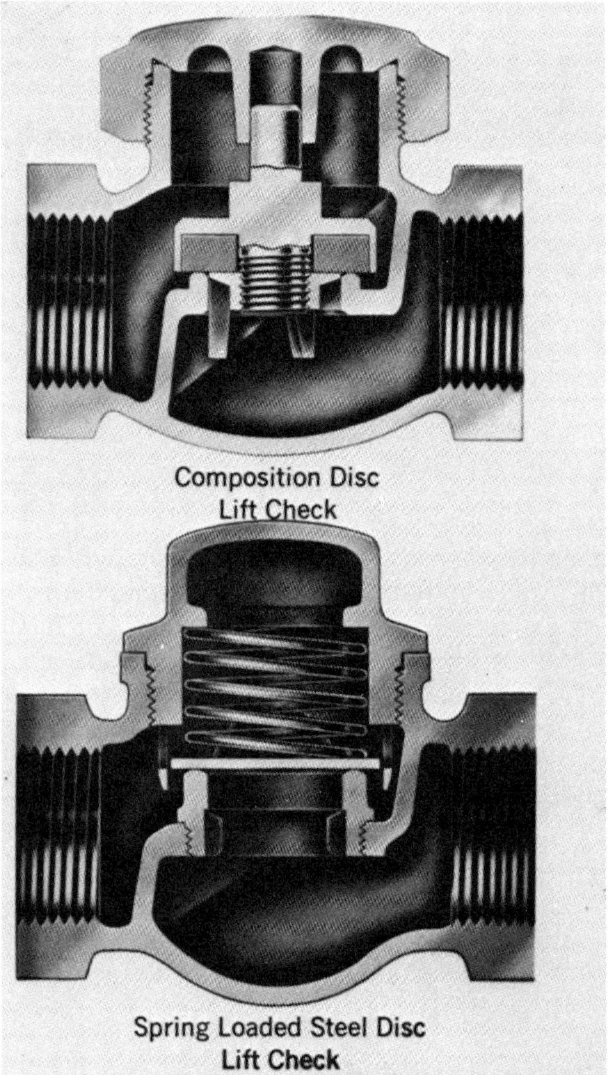

Figure 9-15. Cross section of bronze check valves. (Courtesy of Jenkins Bros.)

whenever the pressure exceeds safe values, and reclose automatically when the pressure drops below the set point of the valve. Pressure reducing valves are designed to maintain a set pressure downstream from the valve when the upstream pressure is considered excessive. These valves can be adjusted, within the operating range of the valve, to provide a constant lower pressure at the valve discharge. A normal application in office-type buildings is in the potable water service. (See Figure 9-16)

Backflow preventers are a special version of check valves. These usually are two check valves in series with an opening to the atmosphere from a center cavity. Backflow preventers are required in all pipelines that provide connection from the potable service to any treated or contaminated water system. A typical application is the isolation of HVAC make-up water. (See Figure 9-16)

Valve actuators may be installed on gate, globe, and ball valves to permit operation

Figure 9-16. Potable water service with parallel pressure reducing valves in the main line and a backflow preventer above in the line supplying water to the HVAC system.

of valves in inaccessible or dangerous locations, and may be either manual or automatic. Manual actuators employ levers, gears, or wheels to facilitate valve movement. Automatic actuators use an external power source to provide the necessary torque to operate the valve to permit remote control or central automated control from where many valves may be interfaced with computers or other automated control systems.

TYPICAL SPECIFICATION
Valves (HVAC)

All valves shall be selected for not less than 125 psi or 150 percent of normal working pressure whichever is greater. All valves for this project shall be the product of one manufacturer when possible. Whenever reference is made to a standard specification, it shall be to the latest issue of that specification.

Valves for the various schedules shall be as follows:

SYSTEM	*SCHEDULE*
Chilled and Hot Water Mains	A
Condenser Water	A
Condensate Drains From Pans	B
Heating, Runouts, Make-up and Fill Lines	B

SCHEDULE A

1. Valves
   Up to $2\frac{1}{2}$-inch shall be quarter-turn ball valves, screwed ends, carbon steel body; 3-inch and larger shall be quarter-turn, threaded lug-type butterfuly valves with carbon steel body and stainless steel disc and memory stops.
   Valves 6-inch and larger to have gear operators.
2. Check valves—swing
   Up to 2-inch shall be bronze body, regrindable, and/or renewable disc, regrindable seat ring, screwed in cap, screwed 125# ends.
   $2\frac{1}{2}$-inch and larger shall be iron body, bronze mounted, regrindable disc and seat ring, bolted cover, flanged 125# ends.
3. Check valves—silent
   All sizes shall be silent spring-loaded type, threaded ends up to 2-inch size, and wafer type $2\frac{1}{2}$-inch and larger.

SCHEDULE B

1. Valves. All sizes shall be quarter-turn ball valves, soldered ends, bronze body, fully ported.
2. Check valves. All sizes shall be swing type, soldered ends. Check valves not specified in valve schedule shall be standard noiseless radial flow-check valves with removable bronze set and interiors designed for 125 psi.

Valve for cooling tower makeup shall be motorized, all brass, two-position, with flanges, designed for 110 V operation.

Butterfly valves (general) shall be a 150# threaded lug type suitable for dead-end service with a carbon steel body and noncorrosive disc, stainless steel shaft. Buna-N or TFE seat suitable for temperatures from $-30°$ to $+225°F$ and an extended neck to clear possible insulation. Valves 5 inches and smaller shall have leverlock handles and valves 6 inches and larger shall be complete with a gear operator and limit switch to indicate open or closed position. Valves shall be furnished with memory stops. Valves shall close bubble-tight in 30 seconds. All automatic operators shall be designed for tight shut-off at full differential pressure. Each valve shall be factor tested.

Valves (Plumbing)

Valves shall be so constructed that they can be repacked under pressure in the open position.

Valves in copper pipe:

1. Gate valves, 2-inch and smaller shall be bronze body, soldered ends, nonrising stem, inside screw, 125 psi. $2\frac{1}{2}$-inch and larger shall be cast-iron body, bolted bonnet, nonrising stem, solid wedge, for 200 psi.

2. Globe valves, 2-inch and smaller shall be bronze body, soldered ends with renewable Teflon disc, 125 psi.

3. Check valves, 2-inch and smaller shall be bronze body, soldered ends, 125 psi; $2\frac{1}{2}$-inch and larger shall be iron body, brass trim, flanged ends, 125 psi.

4. Pressure reducing valves (pressure regulators) shall be the same size as lines in which they are installed, adjustable from 25 to 75 psi, cast bronze with stainless steel fittings, neoprene diaphram, and adjustable screw regulating spring to set discharge

pressure. Provide upstream of the reducing valve an all-bronze type removable strainer. Ahead of the strainer, provide a main shut-off valve and provide pressure gauges on the entering and leaving side of the strainer and on the leaving side of the pressure valve.

5. Provide backflow preventers in all water piping to any system other than domestic water systems in compliance with American Society of Sanitary Engineering Standard.

Valves in steel piping:

1. Gate valves, 2-inch and smaller shall be bronze body, screwed ends, solid wedge, inside screw, rising stem, 125 psi; $2\frac{1}{2}$-inch and larger shall be cast iron body, bolted bonnet, nonrising stem, solid wedge for 200 psi.
2. Globe valves, 2-inch and smaller shall be bronze body, screwed ends, renewable composition disc, 150 psi.
3. Check valves, 2-inch and smaller shall be bronze body, screwed ends, regrindable bronze disc, 125 psi; $2\frac{1}{2}$-inch and larger shall be iron body, brass trim, flanged ends, 125 psi.

Valves (Automation)

Two-way automatic water control valves shall be designed for hot or chilled water service, shall be of packless construction, or with equivalent design features, brass bodies, brass or stainless steel seats and trim. Valves shall be single seated and close tight on shut-off, passing the quantity of water shown on the schedules with a pressure drop through the valve of this flow not more than 15 feet nor less than 5 feet of water. Valves shall have adjustable springs.

Test data showing the actual characteristic of each type and size of valve shall be submitted for approval.

Valves (Fire Protection)

Valves shall be iron body, flanged end, bronze body OS & Y gate valves, UL approved with tamper switches on all valves which control water supply to fire protection system and on post indicator valves. Switches to have normally open and normally closed contacts.

PUMPS AND PIPING

Pumps can generally be classed by performance into two types: positive displacement and variable displacement. Positive displacement pumps maintain a constant delivery against varying discharge pressures, and variable displacement pumps are those whose output varies inversely with discharge pressure. Centrifugal pumps are variable displacement pumps, and as the name inplies, employ centrifugal force to lift liquids from a lower level to a higher level, and are the ones most used in HVAC systems.

A centrifugal pump consists of a vane-carrying wheel impeller and a stationary casing. The liquid enters at the center of the impeller and is accelerated to a high velocity by the rotating impeller. The impeller, through centrifugal action, forces the liquid into the casing and out the discharge port. These pumps produce a constant flow of the liquid without the characteristic pulsations of the reciprocating pumps. However, centrifugal pumps do not maintain constant delivery against varying discharge

Figure 9-17. End suction centrifugal pump. (Coutesy of ITT Fluid Handling Division)

pressures; the flow varies inversely with pressure. The horsepower required to operate the pump varies directly with the flow.

Centrifugal pump discharge can be throttled without building up excessive pressures. But when the discharge is closed completely, the impeller churns the liquid and the input power is dissipated as friction of the water on the impeller. These pumps should not be operated for prolonged periods in this manner as the energy dissipated will create overheating.

Pump performance is best shown by its characteristic curves relating flow, efficiency, horsepower, and head.

Pump curves are arbitrarily classified as *flat* or *steep*. This refers to the general shape of the head-capacity curve. Pump design determines the shape of the curve, and a wide variety of characteristic curves can be obtained. Flat-curve pumps are generally preferred for closed circuit systems because of the influence of the pump curve on the system operating components. Large changes in capacity can be achieved with a small change in head. This is advantageous in balancing multicircuit systems. Steep curve

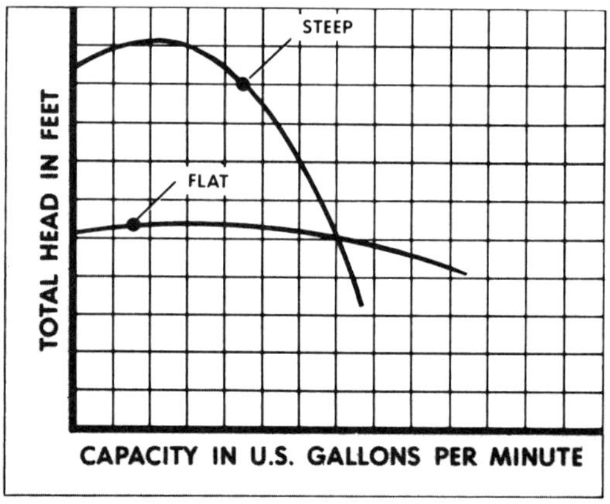

Figure 9-18. Flat vs steep pump curves. (Courtesy of ITT Fluid Handling Division)

pumps are used where relatively constant flow is required with small variations in head. The steep characteristic is obtained by using impellers with backward curved blades.

Selection and Application of Centrifugal Pumps

A knowledge of pump and piping terminology is essential for the proper selection and application of pumps. The pump characteristic curve shows pump performance by means of plotted curves which relate capacity (gallons per minute) to pressure head produced (head in feet of liquid), brake horsepower (power required to drive pump), and efficiency (water horsepower per brake horsepower) to capacity.

The power produced by a pump as it pushes water through the pipe is called water horsepower (*whp*)

$$whp = Q \times H \times S/3960$$

where Q = capacity of pump, gpm
H = total head on pump, feet of water
S = specific gravity of the liquid when liquid is not water.

The power required to drive the pump is called the brake horsepower (bhp) and is determined by actual tests on the pump.

Efficiency of a pump is the output divided by the input

$$eff = whp \times 100/bhp$$

Head on a pump is expressed in feet of water, whereas pressure drop in piping due to friction is expressed in pounds per square inch. These figures are interchangeable as follows:

pound per square inch = 0.433 × foot head of water
or
foot of water = 2.31 × pounds per square inch

The function of a pump is to provide sufficient head or pressure to overcome the losses in the piping system. The total head on a pump consists of one or more of the following:

1. Friction
2. Static
3. Pressure
4. Velocity

There are two types of piping circuits used in chilled water HVAC systems, the closed loop and open piping. Chilled water and hot water systems are closed loop while the condenser water with a cooling tower is an open pipe system.

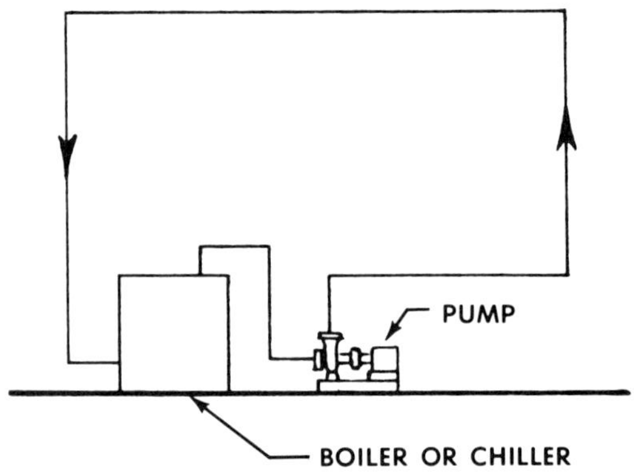

Figure 9-19. Typical closed loop piping system. (Courtesy of ITT Fluid Handling Division)

In a closed loop piping circuit the static and pressure heads are equal to zero, and the only head is that due to friction of the circulating water. The energy head required to raise the water to the top of the supply riser is balanced by the energy head regain as the water flows down the return riser.

Static head or total static is the algebraic sum of the static discharge head and the static suction head. In these cases the total head on a pump consists of the total static head plus friction head loss. See Figure 9–20.

Pressure head (positive or negative, i.e., pressure or vacuum) is developed by the pressure or vacuum of the system in the space from which the water is being drawn or to which it is being pumped. In these cases, the total head on a pump consists of the total static head plus friction head plus pressure head.

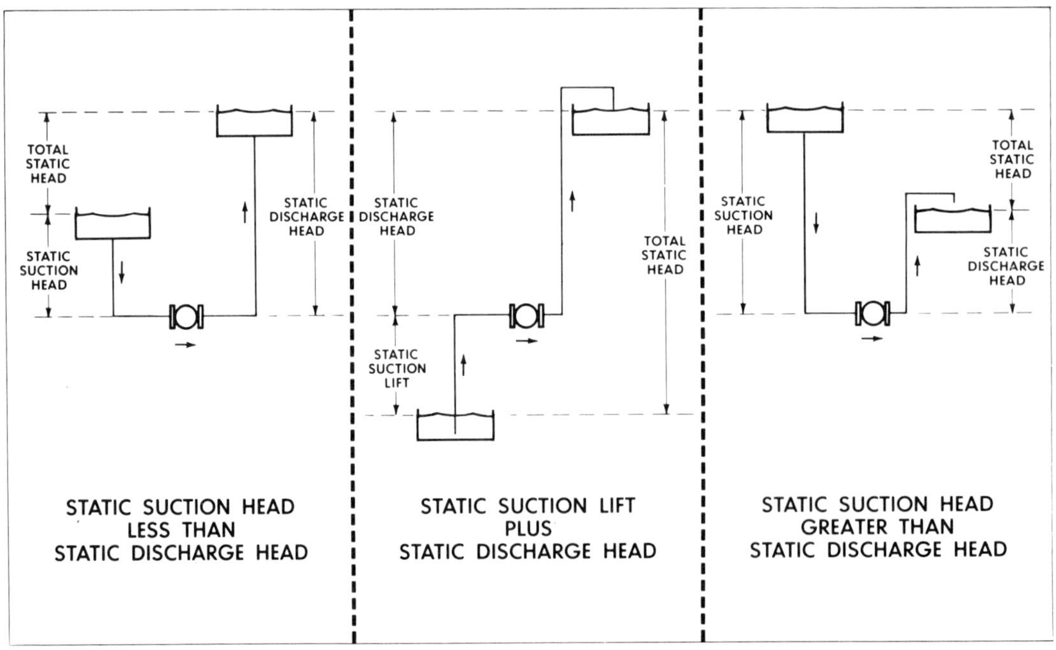

Figure 9-20. Typical open pipe system. (Courtesy of ITT Fluid Handling Division)

The proper pump selection begins with the calculation of the losses in the system. These losses are plotted flow versus pressure drop. Then the required pump operating point is selected. At this point a pump can be selected whose head-capacity curve crosses the system curve at the point of operation. In order to predict pump capability, the system curve should be plotted on the pump characteristic curve. Pump capability follows the head-capacity curve, i.e., the capacity in gallons per minute depends upon the system resistance.

The pump affinity laws provide methods for approximation of the effects of changed impeller diameters and rotational speeds on the pump curve. These laws state:

1. Pump gpm capacity varies directly as the speed or impeller diameter ratio change.
2. Pump head varies directly as the square of the speed or impeller diameter ratio change.
3. Brake horsepower varies directly as the cube of the speed or impeller diameter ratio change.

The relationship of the pump affinity laws can be formulated as follows:

1. $Q_2 = \frac{N_2}{N_1} Q_1, \qquad Q_2 = \frac{D_2}{D_1} Q_1$

2. $H_2 = \left(\frac{N_2}{N_1}\right)^2 \mathrm{H}_1, \qquad \mathrm{H}_2 = \left(\frac{D_2}{D_1}\right)^2 \mathrm{H}_1$

3. $bhp_2 = \left(\frac{N_2}{N_1}\right)^3 bhp_1, \qquad bhp_2 = \left(\frac{D_2}{D_1}\right)^3 bhp_1$

where $Q = gpm$
H = Head
bhp = Brake Horsepower
N = Speed in rpm
D = Impeller diameter

Centrifugal pumps can be operated either in parallel or series. When two pumps are operated in parallel, the head-capacity curve can be plotted by adding the capacity at different heads. This would indicate a flow of the sum of the two pumps when operating against a static head with no friction head. But friction is ever present and the total quantity of flow is always less than the sum of the two pumps. This is best illustrated graphically when the total flow is plotted against the system curve. The system curve also changes, i.e., the head loss due to friction varies approximately as the square of the water velocity.

For series operation the head-capacity curve can be plotted by adding the head at different capacities. For example, where one pump can deliver 100 gpm at a 100 foot head, two identical pumps in series will deliver 100 gpm at a 200 foot head.

The selection of a pump should begin with the design of an economical piping system. Then the pump is selected to match the system curve. While there are many pumps that will satisfy these conditions, the one selected should operate in the area of a maximum efficiency while satisfying system requirements.

Piping systems

The primary consideration in designing a piping system is economy, i.e., the pipe should be large enough to avoid excessive pump operating costs cosistent with reasonable installation first cost.

There are three generally accepted methods for calculating friction loss in the system: constant friction, additive friction, and constant velocity. Of these three, the constant friction method is most commonly used. There are two types of piping systems: closed loop or continuous circuit (circulating service), and open circuit (pumping from and to atmosphere as in cooling tower application).

In the closed loop system, the total pressure drop is due to friction of the water flowing in the pipe, and static and pressure heads are equal to zero. In the open circuit system, the total pressure drop is due to the sum of the friction, static and pressure heads.

Velocity head (head necessary to accelerate the water) is excluded from this discussion because its value is usually so small that it can be omitted. For example: Water flowing at 5.2 gpm in a one inch pipe has a velocity of two feet per second and a pressure drop due to friction of approximately three feet per 100 foot of pipe, but only 0.06 foot velocity head.

It should be considered only when the total head is low and can be calculated by the formula

$$H = V^2/2g$$

where V = velocity in feet per second

g = acceleration due to gravity 32.2 feet per second

and $V = Q/A$

where Q = capacity in cubic feet per second

A = area of the pipe in square feet

Pipe size determines the loss due to friction in any piping system. The larger the pipe, the lower the loss, but pipe costs are higher. The following is a guide used for determining acceptable friction losses in piping design:

Continous duty	1 to 1.5 psi per 100 feet of pipe
Intermittent duty	2 to 2.5 psi per 100 feet of pipe
Occasional duty	3.2 to 3.4 psi per 100 feet of pipe

The fact that pipes are made only in standard sizes materially affects design. Frequently, accurate design calls for in-between sizes, such as 1 $\frac{3}{4}$ inches but, since these sizes are not available, standard sizes must be used. It is the usual practice to use the next larger size rather than to use an undersized pipe; therefore, friction losses through the system will be less than calculated.

Piping design is based upon a straight-forward step-by-step procedure:

1. Determine the flow rate in gallons per minute required.
2. Select the pipe size required from Figure 9–21.

Example: Required flow rate 30 gpm. Select one psi per 100 feet of pipe. Read up from one psi to intersection of 30 gpm and find two inch pipe.

3. Determine total equivalent length of pipe. Add equivalent length in feet for all fittings and equipment, where known, to actual length of pipe. When this information is now known, multiply actual length by 1.6 to obtain total equivalent length. Use Figure 9–22.
4. Calculate total pressure drop due to friction.
5. For closed loop systems construct system curve. Place first point as determined by required flow rate and total head in feet for the system. Additional points can be read directly from a slide rule because pipe friction drop varies as the square of the change in flow rate:

$$\left(\frac{Q_2}{Q_1}\right)^2 = \frac{h_2}{h_1}$$

where:

Q_1 = known (design) flow
Q_2 = final flow
h_1 = known (design) head
h_2 = final head

For a design flow rate of 200 gpm and a head of 30 feet, the other points as shown in Table 9–3 will provide sufficient points so that a system curve can be plotted.

The operation of the pump on the piping circuit described by this system curve *must* be at the intersection of the pump curve with the system curve. This is because of the first law of thermodynamics—energy in must equal energy out. Energy put into the water by the pump must exactly match the energy lost by the water as it flows through

Table 9-3. System Curve Calculations for Closed Circuit Systems.

GPM	FOOT HEAD
115	10
165	20
185	25
200	30
215	35
230	40

(Courtesy of ITT Fluid Handling Division)

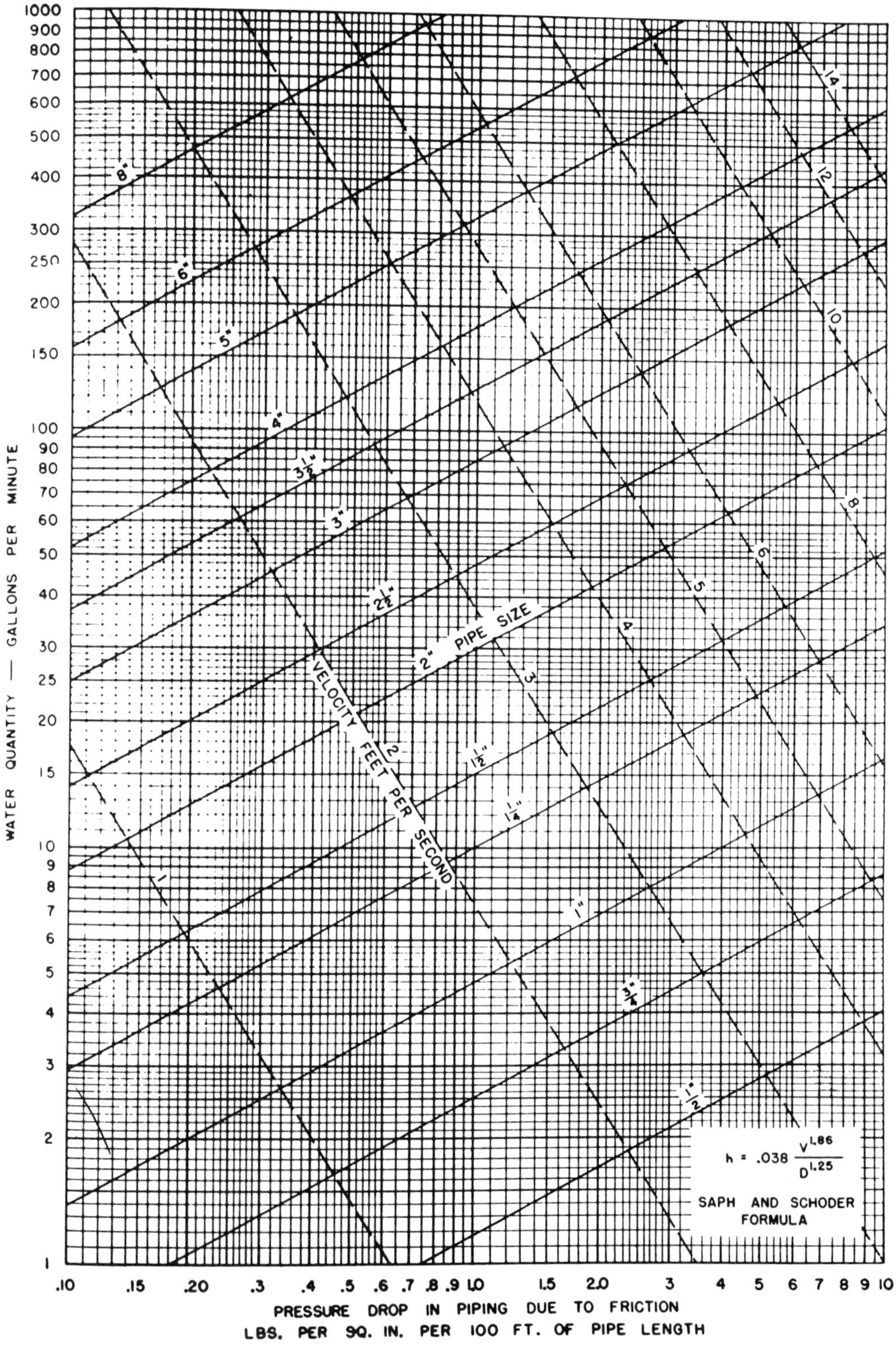

Figure 9-21. Values of pressure loss in this chart are for moderately clean commercial steel pipe. Values of pressure loss in this chart should be increased about 25 percent if the inside of the pipe is rough because of corrosion or scale. (Courtesy of The Trane Company, LaCross, Wisc.)

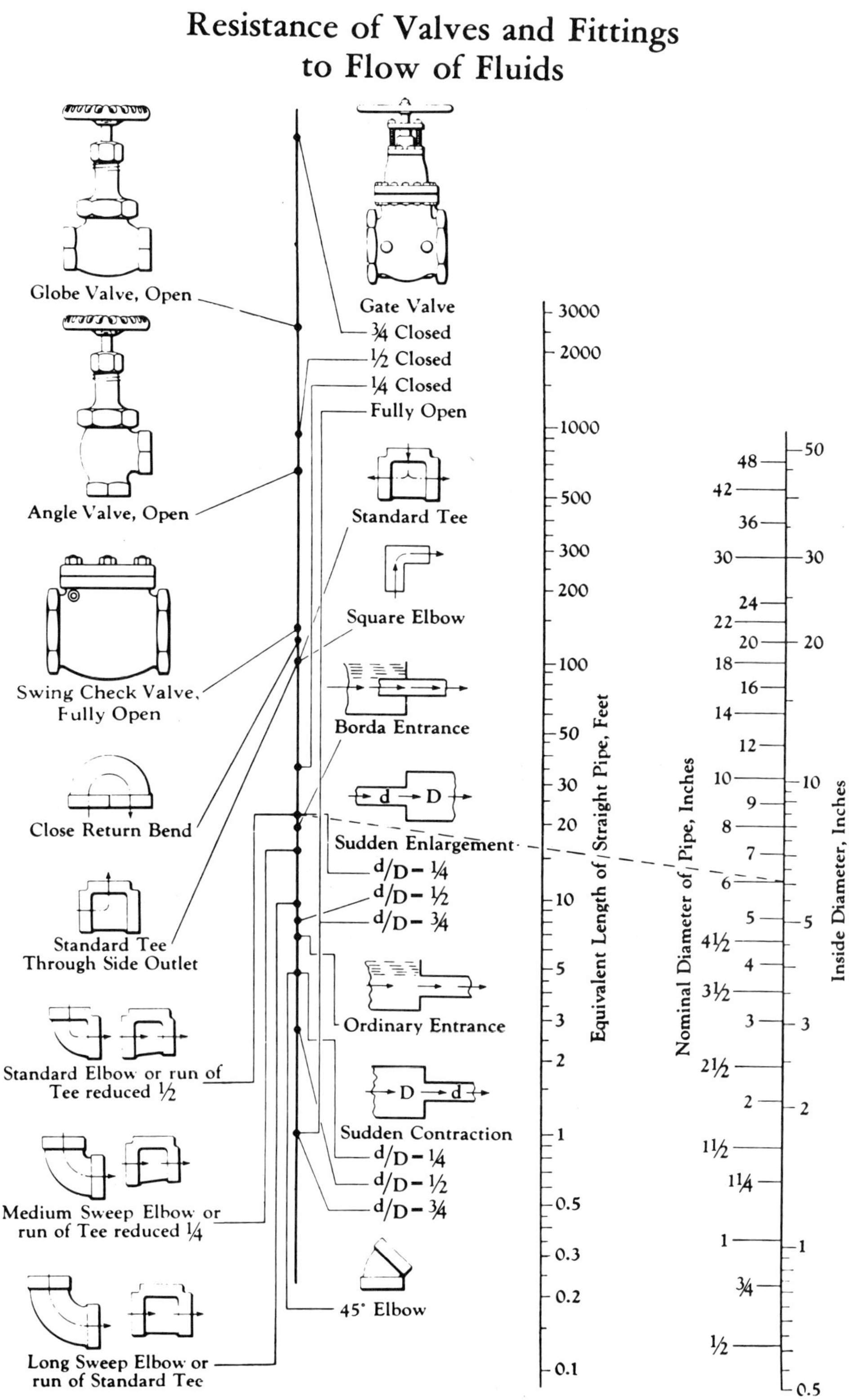

Figure 9-22. (Courtesy of Crane Co.)

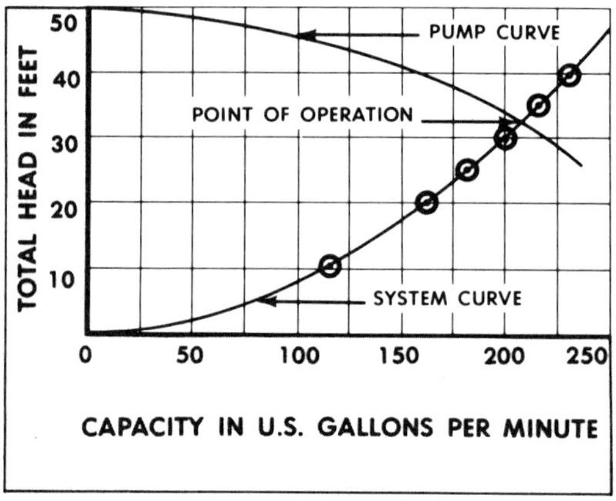

Figure 9-23. System curve plotted on pump curve. (Courtesy of ITT Fluid Handling Division)

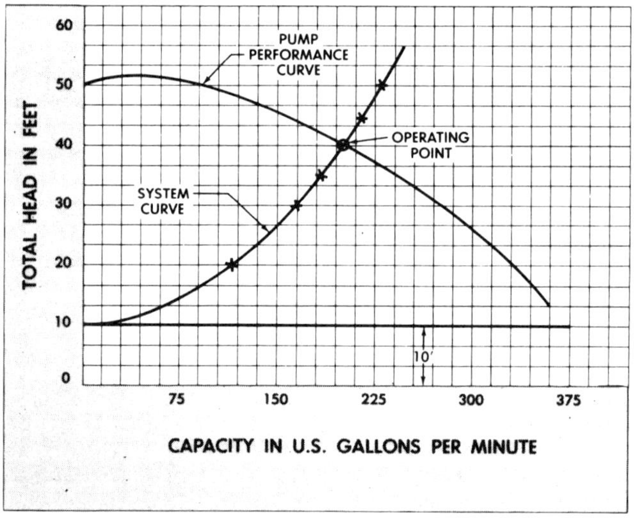

Figure 9-24. System curve for static suction lift. (Courtesy of ITT Fluid Handling Division)

the piping circuit. Water flow through the piping circuit must match pump flow. The point of intersection is the only point that can meet this basic engineering law. For example, the pump will operate at about 210 gpm at 32 feet head, providing the specified selection points 200 gpm and 30 foot head are true statements of the system flow and pressure drop relationship.

6. For open circuit systems the static head loss must be included when constructing a system curve. For the above case the flow rate remains 200 gpm but the total head would become 30 foot friction plus 10 foot static head. Additional points can again be read directly from the slide rule.

A secondary consideration in piping system design is permissible noise level due to water velocity. Too high a velocity will create objectionable noise in the system.

Pump manufacturers develop what are called pump performance characteristics curves under carefully controlled test conditions. Typical pump curve, Figure 9-25, shows pump capacity plotted against total head for different impeller sizes for a specific model pump. Different curves are developed for different pumps. Superim-

Table 9-4. System Curve Calculations for Open Circuit Systems.

GPM	FOOT HEAD
115	20
165	30
185	35
200	40
215	45
230	50

(Courtesy of ITT Fluid Handling Division)

Table 9-5. Acceptable Velocity for Practical Noise Level.

VELOCITY FEET PER SECOND	PIPE SIZE, INCHES INSIDE DIAMETER
$1\frac{1}{2}$	$\frac{1}{2}$
$2\frac{1}{2}$	$1\frac{1}{2}$
3	$1\frac{1}{2}$
4	2
5	3
6	4
7	5
8	6
9	8
10	10

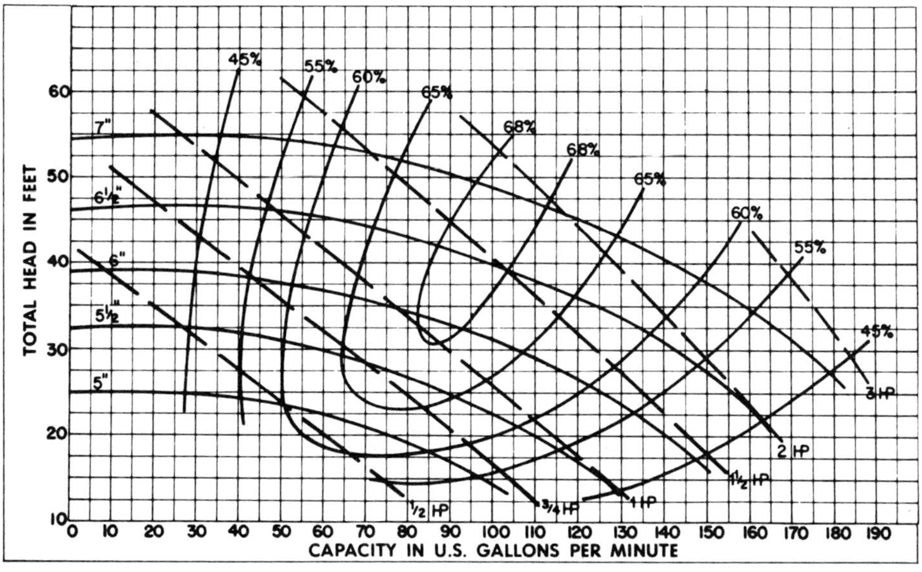

Figure 9-25. Pump curves for 1750 RPM operation. (Courtesy of ITT Fluid Handling Division)

posed on this curve is the horsepower and pump efficiency. After the total head of a piping system is determined, pump selection is a simple process. For example, for a system requiring 100 gpm operating at a total head of 40 feet, the pump will have a 6 $\frac{1}{2}$-inch impeller, 1 $\frac{1}{2}$ horsepower motor, and it will operate at slightly over 68 percent efficiency.

TYPICAL SPECIFICATION
Pumps

1. In-line Pump

Pump shall be close coupled, single stage with a cast-iron body that is bronze fitted. The body shall be provided with a support ring and gauge tappings in both suction and discharge. It shall be possible to service the inside of the pump without breaking any pipe connections. Pump shall be provided with a mechanical seal suitable for the service. Provide a bypass between the seal and the discharge.

The shaft shall be steel fitted with a stainless steel sleeve where it passes through the seal. Sleeve shall be replaceable. The impeller shall be balanced both statically and dynamically and be secured to the shaft. Pump shall be factory tested for performance and sound.

2. End Suction Pump

Pump shall have bronze impeller with vertical split shaft, flanged connection, and be flexibly connected to the motor. It shall have gauge tappings in both suction and discharge. Pump shall be dynamically balanced. It shall be possible to service the pump without breaking any pipe connections. Casing shall be designed for not less than 125 psi and shall deliver the amount of water against the head as listed. Pump shall be provided with mechanical seals suitable for the temperature and pressure encountered and the liquid pumped. Pump shall be furnished with carbon steel shaft provided with stainless steel or monel steel sleeve through the stuffing box, and with bronze split glands with brasss nuts and bolts.

The manufacturer shall guarantee in writing that the pump performance agrees with the published curves.

3. Double Suction Pump

Pump shall have double suction bronze impeller with horizontal or vertical split casing with drain plug, and shall be connected to the motor through flexible metal coupling. Coupling guards conforming to OSHA shall be provided by the manufacturer.

Pump shall be double suction, dynamically balanced, and provided with regreasable self-aligning bearings with an average life of 250,000 hours. An extra strong structural steel base with grout openings shall be provided under the pump and motor. Provide gauge tappings in both suction and discharge. Close grained cast-iron or meehanite casing and flanges shall be designed for 125 psi. Each pump shall deliver the amount of water against the head as listed on the drawings.

Stuffing boxes shall be extra deep and provided with packing suitable for 250°F service. Bronze split glands shall be provided for pump to withstand the operating

Figure 9-26. Installed inline pump with strainer on suction side of pump, pressure gauges on suction and discharge sides, and quarter turn ball valves properly located for servicing purposes.

Figure 9-27. Condenser water centrifugal pump, disconnect switch with capacitor mounted on stand behind switch, butterfly valves, strainer and pressure gauges

Figure 9-28. Same pump. View shows vibration isolators installed between inertia block and equipment pad.

Figure 9-29. Installed packaged air compressor.

Figure 9-30. Installed fuel oil pump package for feed to emergency generator showing appropriate unions, valves gauges.

characteristics of the pump. Pump shall be provided with solid stainless steel shafts or a steel shaft with stainless steel sleeves with 500 BHN (Brinell Hardness), brass nuts and bolts. Manufacturer shall guarantee in writing that the pump performance agrees with published curves. Manufacturer shall submit sound power levels with his pumps.

SUMMARY

Mechanical systems and equipment include HVAC, air distribution, pumps, and valves. Since pumps and valves are used in plumbing and fire protection, that is also addressed.

To understand HVAC, you have to know about the refrigeration cycle and how it works. All refrigeration systems use the same basic principle. Central chiller plants produce chilled water that is pumped to air handlers, where, through an air distribution system heat is extracted from the space. This heat is returned to the chiller where it is transferred through a condenser to condenser water that is piped to cooling towers. The heat extracted from the space is rejected to the atmosphere through cooling towers. The chiller, in order to do its work, contains an air side and a water side. The water side has pipes, pumps, and valves.

1. The refrigeration system uses a fluid (refrigerant) that in the gaseous state is compressed, condensed to a liquid in a condenser, and is changed back to the gaseous state in the evaporator. Heat is extracted through the evaporator and rejected through the condenser.

2. With a central chiller plant chilled water is piped to air handlers.
3. Air handlers use an air distribution system to extract heat from (or during the heating season, may, where necessary, provide heat to) the space.
4. With a central chiller plant this heat is rejected by condenser water through cooling towers.
5. The air distribution system consists of sheet metal ducts (or fiber glass) and terminal devices to distribute air to the space and within the space.
6. The water side of a central chiller uses pipes to contain the water, pumps to move it, and valves to control it.
7. Centrifugal fans are most frequently used with air handling units.
8. Centrifugal pumps are used extensively in HVAC systems and plumbing piping.
9. Globe valves are used for throttling purposes.
10. Gate, ball, and butterfly valves are used for on-off service.

10. Electrical Systems and Equipment

Modern buildings are totally dependent upon the continuous and uninterrupted flow of electrical current. In the event of a power failure, modern buildings soon become uninhabitable. The utility companies have made significant strides to limit the frequency and duration of power failures through a series of complex interconnect systems, and the use of sophisticated control and switching devices. Most in-building distribution systems, equipment, and devices are equally reliable. In the event of a power failure, emergency generators and uninterruptable power sources (UPS) provide temporary back-up power for essential needs. Computers will remain operational until they can be properly shut down or until a short-duration failure is corrected; in medical facilities vital services will continue; and in all modern buildings power will be provided for communications and safe egress.

The state-of-the-art control technology has evolved to a degree of sophistication that almost defies comprehension. Building automation systems are in many instances controlled by microprocessors. But all of this depends entirely upon a very basic power distribution system. While control and current limiting devices are being continually improved, the basic system remains the same. It provides power to the utilization equipment from a source in a safe and efficient manner. A distribution system consists of a series of conductors and overcurrent devices (fuses or circuit breakers) with adequate short-circuit interrupting capacity, switches, transformers, motor control centers, panel boards, controllers, and appropriate metering devices.

POWER DISTRIBUTION

The utility provides power to the site or building at a voltage determined by the requirements of the load and the availability near the site. While the voltage rating is negotiable, the utility usually establishes it. Most utility services, with the exception of that required for residences, are 60-cycle, alternating-current, three-phase, and could vary from 208 volts to 33,200 volts. While voltage ratings may vary somewhat, depending upon local conditions, this section will address nominal voltages.

Outdoor substations are usually installed for use with 33.2 KV and 13.8 KV, and 4160-volt services are terminated at indoor substations. Transformers are the "link" between the high voltage utility service and the building load. Sometimes, depending upon the service voltage, several transformers may be required, in series or parallel, to provide the proper utilization voltage and current to satisfy load requirements. Some of the common voltages are:

Figure 10-1. 33.2 KV outdoor high voltage substation.

Single-phase	
Incandescent lamps	120 V
Fluorescent lamps	277 V
Electric discharge lamps	277 V
Small appliance receptacles	120 V
Motors less than $\frac{1}{2}$ hp	120 V
Three-phase	
Motors $\frac{1}{2}$ hp and larger	208 V
or	480 V
Very large motors	4160 V

The circuit conductors of the distribution system must be encased in some type of raceway, and some special cables are permitted in appropriate cable trays. The National Electric Code permits the use of various types of metallic and nonmetallic raceways depending upon the application.

The circuit conductors are sized to suit the requirements of the load with minimum sizes established by the NEC. These minimum sizes may have to be increased to reduce or limit the voltage drop on long feeder runs. It is very important to maintain proper voltages at the utilization terminals. For example, an induction motor operating at 90 percent of rated voltage will have its starting and maximum torque reduced, its starting and running current increased and its temperature at full load increased about 7° C. With a reduction in voltage, the light output of lamps will decrease and the heat output of resistance heaters will similarly be reduced.

In order to reduce initial capital expenditure for conductors, it is good engineering practice to use the highest practical voltage for the distribution system. For example, at twice the voltage, half the current will be required for the load.

Throughout this chapter typical specification paragraphs will be included, where appropriate, to emphasize the requirements covered in the various sections of this chapter. It is intended to serve as a general guide and not meant to be a complete specification since complete specifications are entirely too voluminous.

TYPICAL SPECIFICATION
Conduit and Wire

a. Conduit
All conductors shall be run in rigid steel hot dip galvanized threaded conduit with threaded couplings and fittings. Branch circuit lighting feeders in dry locations only, above hung ceilings or in nonmasonry partitions may be run in galvanized electrical metallic tubing (EMT). Set screw fittings are not permitted. Rigid heavy wall aluminum conduit may be used in exposed areas and vertical shafts but shall not be used imbedded in concrete.

b. Wire
High voltage 15 KV cable shall be copper single conductor shielded with ungrounded neutral. Cable shall be ozone resistant with insulation composed of thermosetting material derived from filled chemically cross-linked polyethylene. Shielding tape and ground wire shall be copper.

Power and lighting conductors, 600 volts and below shall be copper, 600-volt insulation with a flame and moisture resistant finish of sizes shown on drawings. Minimum size wire shall be No. 12 AWG. Control circuit wiring may be No. 14 AWG. All wires up to No. 10 AWG shall be solid type THW and all wires larger than No. 10 AWG shall be stranded type THWN.

All wires shall be color coded. System voltage 120/208V, Phase A black, Phase B red, Phase C blue, Neutral white. System voltage 277/480 V, Phase A brown, Phase B yellow, Phase C orange, Neutral gray. Grounding conductors green or green with yellow stripes.

TRANSFORMERS

Transformers are devices for transferring energy in an a-c system from one circuit to another, and are used for "stepping up" or "stepping down" the voltage or current. They consist of two or more coils linked by a common magnetic circuit. They are used to step up the voltage at the generator station to a much higher value for economic power transmission, and to step this voltage down for distribution and utilization. They are one of the simplest of all electrical devices; they have no moving parts. Transformers are practically maintenance free when properly installed and operated under normal conditions. Transformers may be divided into three general classes:

1. Instrument
2. Constant-current
3. Constant-potential

This chapter will deal with constant-potential transformers that are used for industrial and commercial light and power service and distribution.

NEC Article 450—*Transformers and Transformer Vaults* provides all the essential details for installation, overcurrent protection, vault requirements, and ventilation.

Since the bulk of the power generated and transmitted is three-phase, these groupings of transformers are of greatest interest. Most frequently, single-phase power is obtained from three-phase connected transformers.

The most common types of connections are:

1. Wye-wye, Y-Y
2. Delta-delta, Δ-Δ
3. Wye-delta, Y-Δ
4. Delta-wye, Δ-Y
5. Open delta, V-V
6. Tee-tee, T-T
7. Scott

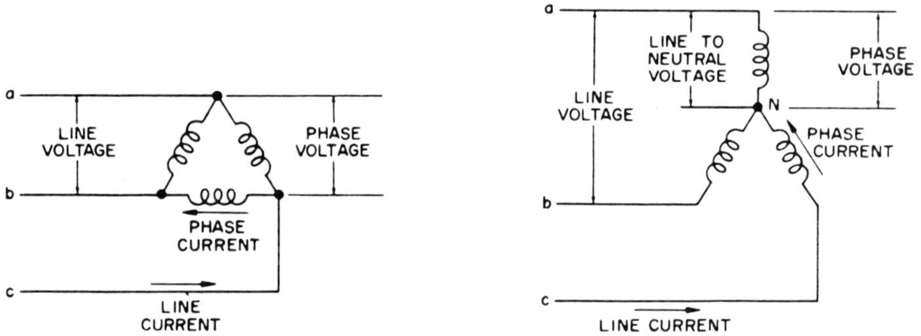

A. Delta-connected. **B. Wye-connected.**

Figure 10-2. Line and phase voltage and current relationship for 3-phase transformer connections.

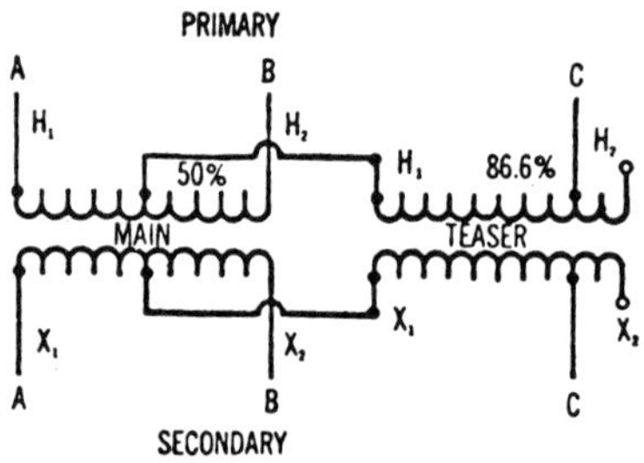

Figure 10-3. Tee-tee connected transformer.

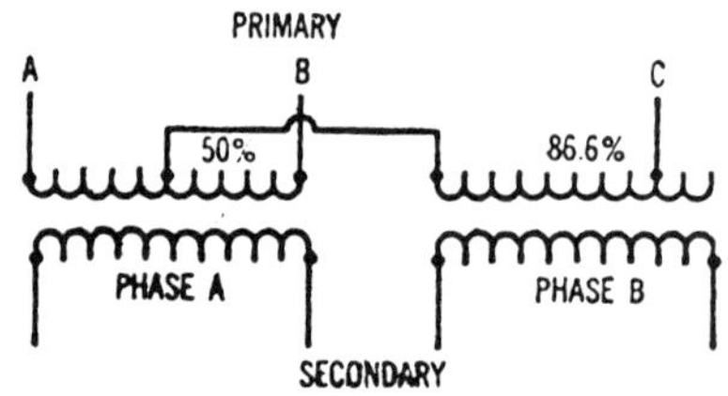

Figure 10-4. Scott connected transformer.

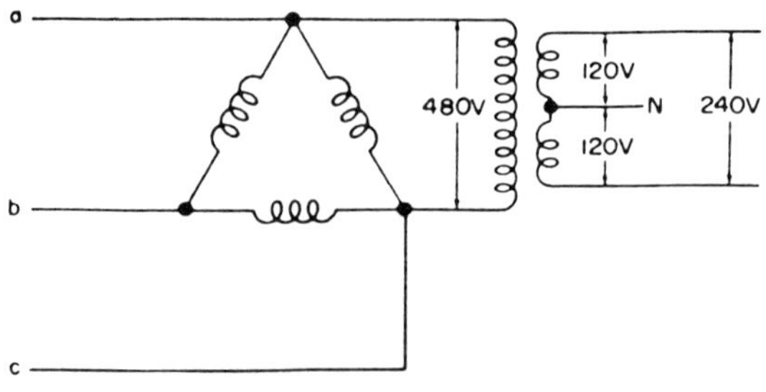

Figure 10-5. Single phase 120/240 volts obtained by connecting a transformer to any 2 lines of a 3-phase circuit with 480 volt rating.

Delta-wye and wye-delta connections are considered the most satisfactory three-phase, three-wire connections. The neutral can be brought out for grounding or for loading when the connection is delta-wye, and for grounding or for connection to the generator neutral when the connection is wye-delta. The neutral is stable, being locked in by the delta. The required third-harmonic component of the magnetizing current circulates in the delta, and no third-harmonic component appears in the phase voltage on the wye side.

The kva rating of a three-phase transformer is the same whether delta- or wye-connected:

$$kva = \sqrt{3} \times VI/1{,}000$$

Where V = Voltage between phases
I = Current in amperes in any phase

The difference is in the method these voltages and currents are obtained:

Wye-connected:
line current = phase current
line voltage = $\sqrt{3}$ × phase voltage

Delta-connected:
line voltage = phase voltage
line current = $\sqrt{3}$ × phase current

From this it becomes obvious where some of the earlier mentioned voltages are obtained. In order to obtain the necessary 120 volt single-phase power from three-phase wye-connected transformers, the three-phase power becomes 208 volts, i.e., $208 = \sqrt{3} \times 120$. The 120 volts are available from any phase to neutral. In the case of 277 volts now commonly used for fluorescent lighting, this is obtained from a 480 volt three-phase wye-connected transformer, i.e., $480 = \sqrt{3} \times 277$.

When the secondary is delta-connected (this is known as the ungrounded system and is used in many industries) single-phase transformers are required to obtain the necessary 120 volt power.

NEC Section 215-8 permits the midpoint of one phase of a delta connected transformer to be grounded to supply lighting and similar single-phase loads providing certain conditions are satisfied; one being that the high (voltage) leg be identified by an orange colored conductor. See that section for other provisions.

The only real advantage of an open delta-connection is that it requires but two single-phase transformers. This permits voltage transformation from one three-wire, three-phase circuit to another. When one unit of a delta-delta bank is disabled, the open delta is an emergency expedient. The maximum kva that can be delivered is 57.7 percent of the kva of the delta-delta bank if the transformers are not to be overloaded.

The tee-tee connection permits the transformation of power at a certain voltage in a given three-wire, three-phase circuit to power at another voltage in another three-wire, three-phase circuit by the use of two transformers. The *main* winding has a tap at midpoint on both the primary and secondary sides. The *teaser* has but 86.6 percent of the turns of the main. In most cases, the transformers used for this connection are identical—each has a 50 percent and an 86.6 percent tap. This connection is usually compared with the open delta, as each requires two transformers. It has numerous advantages over the open delta, but neither is considered first choice in any usual installation.

Transformation from three-phase to two-phase is accomplished by the Scott connection. The three-phase side of the transformer is connected as outlined above for the tee-tee transformer. On the two-phase side, 100 percent taps are used, and the two transformer leads are kept separated and form the necessary two-phases. As an example, the two single-phase transformers are 50 kva each and the transformation ratio is one to one, i.e., 220 volts to 220 volts. Therefore:

$$\text{Two-phase I/phase} = \frac{50 \text{ kva} \times 1{,}000}{220 \text{ volts}} = 227 \text{ amperes}$$

$$\text{Three-phase I/phase} = \frac{100 \text{ kva} \times 1{,}000}{\sqrt{3} \times 220 \text{ volts}} = 262 \text{ amperes}$$

or 15.5 percent greater than rated current will flow in the transformers. From this it is obvious that the windings on the three-phase side of a Scott-connected pair of transformers must be designed to carry 15.5 percent more current than the single-phase ratings of the transformers. In other words, the rating of the bank should be reduced to 86.6 percent of the combined single-phase ratings if no rated currents are to be exceeded.

In both the tee-tee and Scott connections similar transformers can be used. These can have 50 and 86.6 percent taps on both the primary and secondary, or one can have 50 percent tap and the other can have 86.6 percent. Economics would dictate the particular choice. In either case, there is a very definite advantage in the proper use of these connections.

The Scott connected transformers are necessary in an area that was supplied by two-phase power but is being changed to three-phase distribution. These transformers should be so selected that, when the equipment within the plant is converted to three-phase, merely reconnecting the transformers from the Scott to the tee-tee connection will supply three-phase to the plant. It is in this area that the tee-tee connected transformers see the greatest application.

Transformers can be generally classed by cooling method and insulating material:

1. Liquid-filled (liquid-cooled)
    a. mineral oil
    b. silicone fluid
2. Dry-type (air-cooled)
    a. ventilated
    b. non-ventilated
3. Cast-coil, dry-type

4. Gas-cooled
5. Vaporization-cooled

The most frequently used types are the liquid-filled and the dry-types. The others are used to a lesser extent and usually find application in special cases. These are the gas-cooled, vaporization-cooled and the recently developed cast-coil, dry-type. One type of liquid-filled transformer uses a mineral oil as the insulating material and the other uses a silicone fluid for nonflammable applications for PCB fluids.

Oil-immersed transformers have been most widely used since the a-c distribution system was developed. These have the advantage of lowest first cost and inherently weather proof construction. To prevent deterioration of the oil and insulation, they must be maintained pressure tight. The installation of oil-immersed transformers must comply with the requirements of the National Electric Code, as special fire resistant enclosures are necessary when installed indoors.

The dry-type transformer was developed to fulfill the need for indoor instalations where the use of liquid-immersed transformers creates problems. It does not have the liquid maintenance and other problems of liquid immersed transformers. The initial cost of the dry-type is about the same as for the silicone-filled, but its total weight is less than either type of liquid immersed transformer. This may be an important consideration in the total installation cost.

Specific provisions applicable to different types of transformers are outlined in the National Electric Code, article 450.

The transformer selection problem is essentially based on safety and economics. First, consideration should be given to location, as the most economical location is where the transformer is as near the load center as possible. Then the correct type of transformer should be applied, giving consideration to any peculiar problems.

A selection and installation checklist will serve in the decision making process:

A. Location
   1. Indoor
      a. Vault
      b. Enclosure
      c. Open
   2. Outdoors
      a. Enclosure
      b. Shelter
B. Type
   1. Legend-filled
   2. Dry-type
C. Initial Cost
   1. Transformer
   2. Vault
   3. Enclosure
   4. Shelter
D. Maintenance

The choice for outdoor installations is usually the oil-insulated transformer, but the dry-type, non-ventilated is prefered if a fire hazard exists. For indoor installations the dry-types are usually first choice

Dry type transformers are ideally suited for indoor application and offer many advantages over liquid filled transformers. They may be installed in practically any indoor location not subject to submersion or to a high concentration of destructive fumes.

Air insulated and cooled by natural convection (or fan cooled) of air, they are safe and cannot explode, no toxic gases can be released and fire hazards are negligible.

TYPICAL SPECIFICATION
Dry-type Transformer

The open ventilated dry-type transformer shall be provided with four approximately $2\frac{1}{2}$ percent rated kva taps in the high voltage winding, two above and two below rated primary voltage brought out to an internal terminal linkage, accessible through a removable section. The transformer base construction shall be of the fabricated type and suitable for using rollers or skidding in any direction. The transformer shall consist of the following:

Self cooled/fan cooled rating	2500/3333 kva
Temperature rise	150°C
Frequency	60 Hertz
Number of phases	three
High voltage rating-delta	13200 volts
Low voltage rating-wye	480Y/277 volts
Impedence	5.75%
Sound level	64 decibels

Double-ended substations are commonly used in lieu of separate transformers and switchboards. These substations usually contain two transformers, one for 480Y/277 volts and the other 208Y/120 volt distribution, high voltage switching and fuses, lightning arresters, low voltage distribution circuit breakers, current transformers, ammeter, voltmeter, and watthour meter in one manufactured and factory assembled and wired unit. Other than the high voltage cable termination to a common bus, no field wiring is required.

MOTORS

The ruggedness and simplicity of three-phase squirrel-cage induction makes them first choice where constant speed is acceptable. For this reason, they comprise about 90 percent of all electric motors currently in use. An induction motor is simply an electric transformer whose secondary is separated by an air gap and is permitted to rotate. It consists of a few pounds of copper, some special iron of different types, and insulation put together in accordance with a definite but relatively simple electrical law in such a manner as to produce rotative power.

The rotor consists of circular laminations with slots on the outer surface slightly askew to the axis. Rotor conductors, or bars, of copper, brass, or aluminum are

Figure 10-6. Secondary substation with double ended distribution switchboards.

pressed or cast into the slots. All these conductors are mechanically and electrically connected by end rings. These conductors and end rings form a structure that resembles a squirrel's cage, hence the name.

The stator consists of a stack of circular laminations with slots on the inner surface parallel to the axis and attached to a yoke or frame for support. The primary winding is placed in these slots and it is the insulation on or around this winding that determines the life of the motor.

The most vulnerable part of the motor is the stator winding insulation. This insulation is subject to thermal aging. Insulation seldom fails from immediate breakdown upon arrival at some critical termperature, but as a result of gradual deterioration of the dielectric capability with time. There is a specific temperature rise that is permitted by standards, based upon the insulating material properties, and for every 10°C rise above that limit, insulation life is halved.

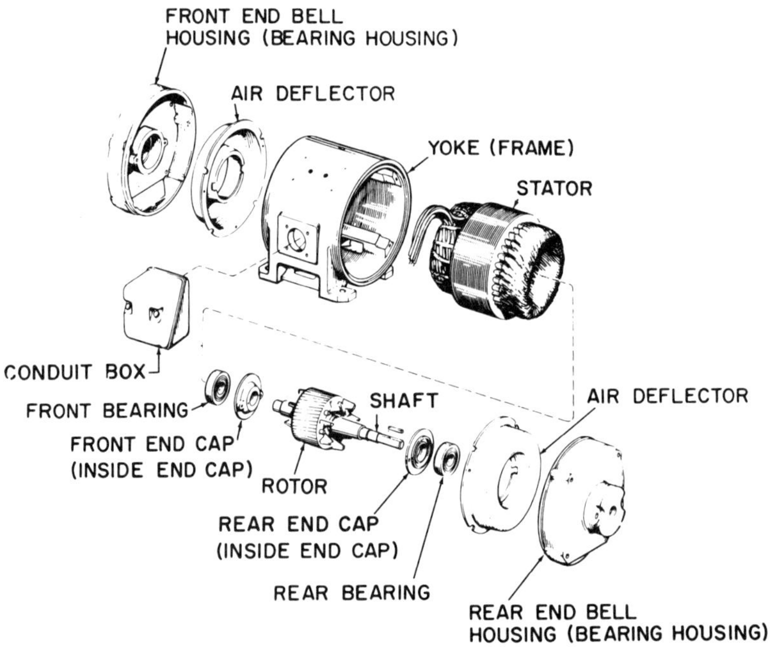

Figure 10-7. Exploded view of a 3-phase motor with parts nomenclature. (Courtesy of Siemins-Allis, Inc.)

Figure 10-8. Cut-away view of a 3-phase squirrel-cage induction motor. (Courtesy of Gould Inc., Electric Motor Division.)

Many insulation failures are the result of a lack of thorough understanding of the temperatures for which motors are designed. The temperature at which a motor operates normally is determined by the insulation used in accordance with the following classifications:

CLASS	HOTTEST SPOT TEMPERATURE
A	105° C
B	130° C
F	155° C
H	180° C

Ever hear anyone complain, after placing his hand on a motor, that it was "too hot?" It most likely is too hot, to the touch. Safety engineers set the limit for human flesh at about 150° F (65.5° C). From this, it is obvious that some parts of a fully loaded motor will be uncomfortably hot to the touch, even motors with Class A insulation.

The second most vulnerable part of a motor is the rotor bearing. Roller contact ball bearings are available in two general types: the open bearing requiring field lubrication; and the sealed, prelubricated type which cannot be lubricated in the field. Each has its place in industry and the selection depends upon the specific application. In inaccessible and out-of-the-way places, where lubrication is at best hit or miss, the sealed bearings are first choice.

Induction motors are available in a number of different frame types, some of the more common are listed:

1. Open
2. Drip-proof
3. Splash-proof
4. Weather-protected
5. Totally-enclosed
6. Totally-enclosed fan-cooled
7. Explosion-proof

The most common, and also the least expensive, of the group are the open-frame motors. They are best suited for clean indoor areas. These motors draw air through vent openings over and around the windings for cooling purposes. While a certain amount of normal dirt can be tolerated, an excessive amount will restrict ventilation and will cause insulation failure due to excessive temperatures.

In cases where moisture conditions are encountered, the drip-proof, splash-proof, and weather-protected motors are needed, depending upon the severity of the conditions. Many plants use the totally-enclosed fan-cooled (TEFC) motors exclusively because of their lower maintenance costs, increased reliability, and maximum flexibility; and also because these motors can be used interchangeably in most locations, thereby requiring fewer spares. However, the higher first cost often offsets this advantage. Most companies, for economy reasons, usually use several different types. The explosion-proof motor is for hazardous locations and is considered a special application case.

Motor selection requires matching the motor characteristics to the requirements as follows:

Load requirements
1. Torque
2. Horsepower
3. Speed

Ambient Conditions
4. Enclosure
5. Bearings
6. Bearing lubrication
7. Insulation

Power Source
8. Voltage
9. Frequency
10. Phase

While most of the above characteristics are readily matched with the requirements, torque usually creates problems as it is less understood and also because induction motors have four values associated with them:

1. Starting torque or locked rotor torque is that torque which a motor develops at rest.
2. Pull-up torque is the minimum torque developed by a motor during acceleration.
3. Breakdown torque is the maximum torque a motor will develop without an abrupt drop in speed.
4. Full-load torque is the torque necessary to produce rated horsepower at full-load speed in pound feet:

$$\text{Torque (full load)} = \frac{5{,}250 \times \text{Horsepower}}{\text{Rpm (full load)}}$$

To satisfy the different torque needs of industry, motors are manufactured in four design classifications: A, B, C, and D. Torque characteristics depend upon rotor-cage design, and three basic designs cover the range; i.e., single-cage, double-cage, and deep-bar rotor. These different rotor designs also affect starting current and, therefore, many possible combinations of starting current, and starting and maximum torque are available for varied applications.

Design B is for general-purpose across-the-line starting duty and is the most commonly used. It has a reasonably high starting torque for accelerating high-inertia loads and can handle short-duration overloads before reaching the breakdown point. Where the load duty-cycle has a peak in access of the Design B breakdown torque a Design A motor may be used.

The amount of current that will be drawn by a motor under locked-rotor conditions

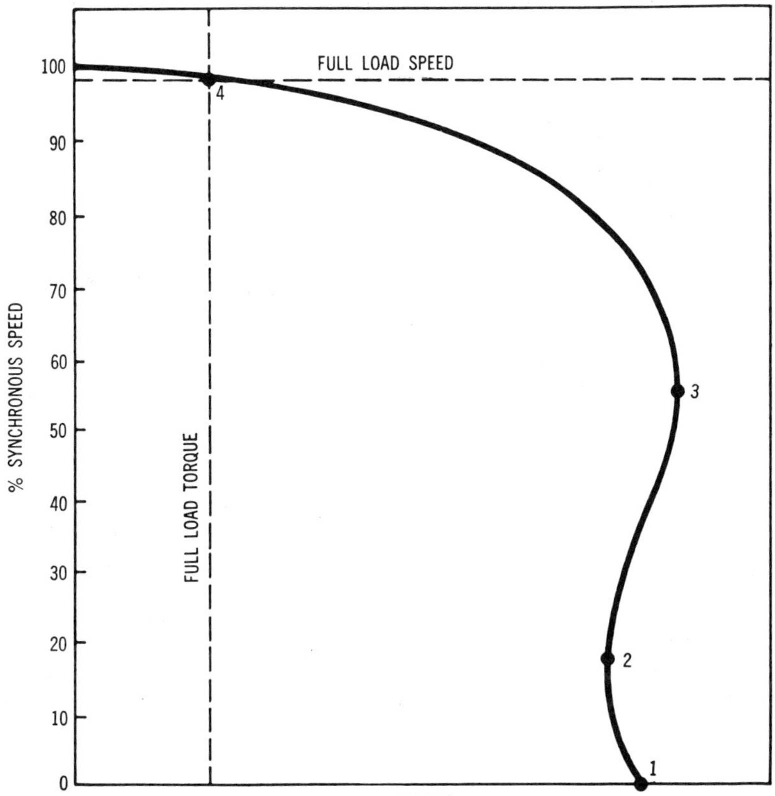

Figure 10-9. Typical Speed-Torque Characteristics of a Squirrel-cage Motor from zero to full speed. Motor Torque defined at four points: 1. Locked Rotor or Starting Torque is the minimum torque a motor will develop at rest for all angular positions of the rotor with rated voltage applied at rated frequency. 2. Pull-up or Minimum Torque. 3. Breakdown Torque is the maximum torque a motor will develop with rated voltage applied at rated frequency without an abrupt drop in speed. 4. Full Load Torque is the torque necessary to produce rated horsepower at full speed.

is identified on the motor name plate by a code designation as "locked-rotor kva." The formula to determine locked-rotor current is as follows for 3-phase motors:

$$\text{Locked-rotor amps} = \frac{\text{kva/hp} \times \text{hp} \times 1000}{1.732 \times \text{line voltage}}$$

This current, according to NEC, "shall be used for determining branch-circuit overcurrent protection. . . ." The NEC table 430–7(b), Locked-rotor indicating Code Letters lists letters from A through V, for example A = 0–3.14 and V = 22.4 and up.

The starting current of squirrel-cage induction motors varies from 4 to 10 times full-load current, depending upon motor design. This high value usually lasts only a few seconds (three to five) and then returns to the full-load value. During this period of time the motor attains full-load speed.

A general purpose motor, (open type) or any motor having a rated temperature rise of 40°C continuous (ambient temperature not exceeding 40°C) when operated at

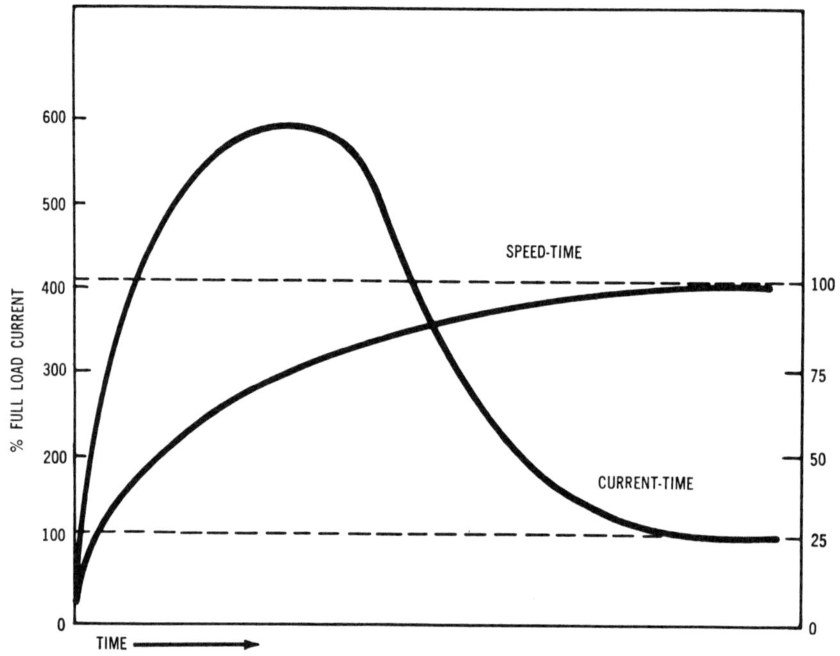

Figure 10-10. Typical Speed-Time, Current-Time Curves for a Squirrel-Cage Motor. Current-time curve shows typical inrush current—this varies from 4 to 10 times full load current, depending upon motor design.

rated voltage and frequency may be overloaded up to the horsepower obtained by multiplying the rated horsepower by the service factor shown on the name plate. The service factors for general purpose, drip-proof, ac, 60 cycle induction motors are:

MOTORS	SERVICE FACTORS
$\frac{1}{6}$ thru $\frac{1}{3}$ hp	1.35
$\frac{1}{2}$ thru $\frac{3}{4}$ hp	1.25
1 thru 200 hp	1.15

The service factor is a built-in "safety factor" which permits the motor (when necessary) to deliver greater than rated horsepower. For example, a motor with a service factor of 1.15 can deliver 15 percent greater than rated horsepower without injurious heating.

Motors are usually wound to permit connection to two voltages. For example, single phase motors are so wound that sufficient leads are brought to the terminal box to permit connection to either 115 volts or 230 volts. These are usually six-lead motors whereby the internal coils may be connected either in parallel for 115 volts or in series for 230 volts. The motor will perform with exactly the same characteristics regardless of the external connection.

Similar connection possibilities are provided for three-phase motors. In these cases the motors usually have nine leads brought to the terminal box. Again, the motor coils may be connected either parallel for 208–230 volts, or in series for 460 volts.

There are several important formulas and conversions for use with three-phase squirrel-cage induction motors. Full-load torque has already been defined.

1 horsepower = 745.7 watts
= 33,000 foot-pounds per minute
= 550 foot-pounds per second
= 42.43 British thermal units per minute (Btu/min)
horsepower output = hp × 0.745.7
horsepower input = hp × 0.745.7/eff
= $\sqrt{3}$ × amperes × voltage × power factor

Another common term applied to motors is *apparent efficiency*.

Apparent Efficiency = Efficiency × Power Factor

The motor name plate is the source for all pertinent motor data. It contains all the information discussed above. For dual voltage motors, it even shows the connection diagram, identifies the lead numbering system, and shows which lead is to be connected where for either voltage. Become familiar with motor name plates.

Motor specifications are usually included in the mechanical division of the specifications because motors usually are provided with mechanical equipment to drive pumps, fans, etc.

TYPICAL SPECIFICATIONS
Motors

This contractor shall provide and install all motors for equipment he purchases.

All constant speed motors shall be squirrel-cage, designed for normal starting torque and normal starting current, in accordance with NEMA standards, and have a service factor of 1.15. Motors shall be provided with ball bearings and grease fittings.

Provide power factor correction capacitors to obtain 95 percent power factor at full load. Capacitors shall be furnished in an enclosure, and loose for field mounting and wiring under Division 16.

Motors $\frac{1}{3}$ hp and smaller shall be 115 volts, 1 phase, 60 Hz.

Motors $\frac{1}{2}$ hp and larger shall be 460 volts, 3 phase, 60 Hz.

Motors shall have Class B insulation.

Motors 1 thru 25 hp shall be high efficiency types.

The apparent efficiency shall meet or exceed the values in the following table when tested in accordance with IEEE test procedure 112A, Method B:

HP	APPARENT EFFICIENCY
1	66.9%
$1\frac{1}{2}$	69.7
2	70.6
3	71.9
5	73.4
$7\frac{1}{2}$	74.4
10	75.0
15	75.9
20	76.5
25	67.8

SHORT-CIRCUIT CALCULATIONS

Electrical building distribution systems are designed to carry power to utilization equipment such as motors, resistance heaters, lighting, etc. A well-designed system should be capable of providing power under normal operating conditions and prevent damage to the system and equipment, and provide for the safety of people, during abnormal conditions.

A good example is the combination motor starter. It has overcurrent relays which are responsive to motor overloads up to 8–10 times full-load currents of the motor. This eliminates nuisance tripping on start-up of the motor and during momentary overloads. Above 8–10 times full-load currents, the overcurrent enters the short-circuit region, requiring a secondary protective device to be added to the starter. These two devices give the starter a "combination" of overload and short-circuit protection.

This short-circuit protection is provided by fuses, circuit breakers, or circuit breakers with fuses on the line side of the circuit breaker to protect it against fault currents beyond its design limit.

The interrupting capacity of these devices must be adequate to handle fault currents which can be many times the normal load currents.

Unfortunately, there is no simple clamp-on ammeter that will show the maximum current that can flow through a fault. This must be calculated—a very complicated task if the value is to be determined with great precision. Rigorous determination of short-circuit currents requires accurate reactance and resistance data for each power carrying component from the utility generating station right to the point of the fault.

The following study is a simplified method for calculating fault currents on three-phase systems in which there are more than one circuits at different voltages connected by transformers.

Utilities can and do furnish data on the available kva at the primary terminals of their transformer (or owners, depending upon local conditions). The largest possible source is called an "infinite bus." A 500 mva, primary source is generally considered the largest that will be encountered in a practical system. (1 mva = 1000 kva = 1,000,000 VA) Any source greater than 500 mva is considered unlimited. Some studies have been based on 1000 mva. This short-circuit study uses 750 mva as available from the utility.

Possible sources of short-circuit currents, listed in their order of importance or likelihood of being present:

1. Utility systems
2. Induction motors
3. Generators
4. Synchronous motors

The amount of short-circuit currents delivered by a transformer is dependent upon the following:

1. Primary source (available KVA)
2. Secondary voltage

3. Size or rating of transformer
4. Impedance of the transformer

Fault current is limited by the impedance of the circuit between the source or sources and the fault, and is directly related to the size or capacity of the power source. In this case, the two sources are the utility company and the induction motors.

After a fault occurs, inertia continues to drive a motor, causing it to generate current. This drops off rapidly, disappearing entirely after a few cycles. Amperage is approximately equal to the full voltage starting current of the motor, and may be estimated at 3.6 times full load current of all connected and running induction motors. Where the connected motor load is unknown, the following approximations should be used. Based on motor load distribution for 120/208 v, and 240 v, 480 v, and 600 v systems, the total motor short circuit current (I_{scm}) should be added to the calculated system short circuit current as follows:

120/208 v system: $I_{scm} = 2\ I_t$
240 to 600 v system: $I_{scm} = 4\ I_t$
where I_t = system full load current

In some applications, a short can occur where motor load will not contribute energy. In such cases, of course, motor contribution current can be ignored.

Generators, either in the plant or on the utility system, provide symmetrical short circuit current. Since a generator continues to be driven and to have its field energized by a separate exciter, the steady state value of the fault current will persist until interrupted.

Whether the fault current is symmetrical or asymmetrical depends upon what point on the voltage wave the fault occurs. If it occurs at or near a voltage peak, the fault current wave starts at zero and is symmetrical about the zero axis.

If the fault occurs at or near zero on the voltage wave, the fault current wave again starts at zero on the original zero axis. However, at the start of the fault the current should be at or near peak since it lags voltage by 90° and voltage is at zero.

Both are substantially sine curves and are related in phase angle by the impedance angle of the system to the point of fault. Since resistance will usually be negligible in comparison to reactance, the fault current will usually lag the source voltage by nearly 90°.

The result is that the fault current wave becomes symmetrical about a new zero axis, and is offset or asymmetrical with respect to the zero axis of the voltage wave. This new axis is known as the d-c component axis. Magnitude of the d-c component can vary from zero to a maximum equalling the peak of the a-c component. A multiplying factor of 1.25 has been established as representative of conditions $\frac{1}{2}$ cycle after fault occurrence, and should be used to convert a-c symmetrical current to the total asymmetrical current. The d-c component decays to zero and usually disappears after a few cycles. Decay rate is proportional to the ratio of reactance to resistance (X/R ratio) of the system from source to fault.

A simple step-by-step procedure provides all the necessary information required to calculate available short circuit currents for any distribution system:

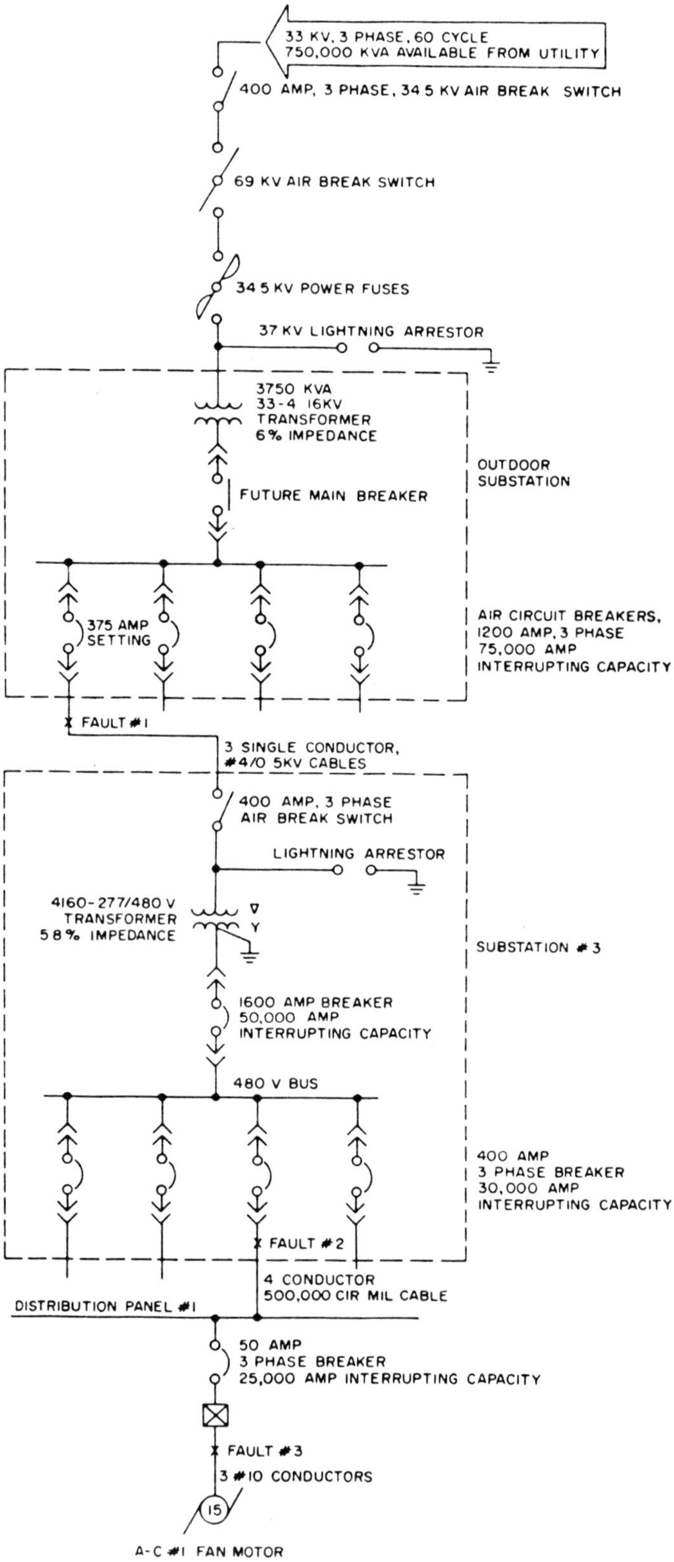

Figure 10-11. One-line diagram of electrical distribution system is first step in short circuit study.

1. Prepare a one-line diagram showing all sources of short-circuit currents and all significant circuit elements, or use the one-line diagram made by the consultant. Where similar distribution (length and size of conductors) runs exist, usually any one may be analyzed.
2. Prepare an impedance diagram so that all calculations can be properly displayed.
3. Obtain available short-circuit mva from the utility.
4. Calculate all appropriate per unit impedance values.
5. Calculate available short-circuit currents.

The best method for presentation of short-circuit calculations is by a case study. The first step in making a short-circuit study is to prepare a one-line diagram showing all sources of short-circuit currents and all significant circuit elements. From this, draw a simple one-line impedance diagram using all significant reactances and resistances—recognizing, of course, that only reactance will be used in many cases.

Impedances may be expressed in ohms, per cent, or per unit. The per unit system (symbol is °/1) is a means of expressing numbers for ease of comparison. Expressed as a decimal, a per unit value is the ratio of a number to a base number. The relationships are:

$$\text{Impedance in ohms:} = \frac{(\%\,Z)\,(kv^2)\,(10)}{\text{base kva}}$$

$$\text{Per cent impedance} = \frac{(\text{ohms } Z)\,(\text{base kva})}{(kv^2)\,(10)}$$

$$\text{Per unit impedance} = \frac{(\text{ohms } Z)\,(\text{base kva})}{(kv^2)\,(1{,}000)}$$

$$\text{Per unit impedance} = \frac{\%\,Z}{100}$$

Where: Z = impedance to neutral
Kv = line to line base voltage in kv
Base kva = three-phase base kva

Per unit quantities can be used to great advantage in determining current, voltage, etc. throughout a three-phase system with many circuits connected by transformers operating at different voltages.

After determining the system and equipment impedances, combined them into a single equivalent impedance from the source to the fault. From this you can determine the value of symmetrical short-circuit current and kva as follows:

$$\text{Short-circuit current} = \frac{\text{base kva}}{(°/1Z)\,(\sqrt{3})\,(kv)}$$

$$\text{Short-circuit kva} = \frac{\text{base kva}}{(°/1Z)}$$

Where: kv = line to line kv

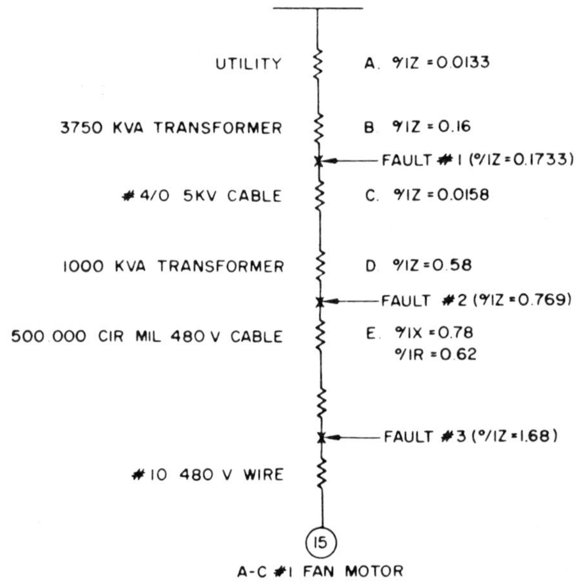

Figure 10-12. Impedance diagram for the system shown in Figure 10-11.

For simplification, and to yield practical numbers, a base kva of 10,000 is selected. To convert ohms to per unit values, a multiplier is established as follows:

at 4,160 V

$$\text{Per unit impedance} = \frac{(\text{ohms } Z)\ (10{,}000 \text{ kva})}{(4.16)^2\ (1{,}000)} = (0.576)\ (\text{ohms } Z)$$

at 480 V

$$\text{Per unit impedance} = \frac{(\text{ohms } Z)\ (10{,}000 \text{ kva})}{(0.48)^2\ (1{,}000)} = (43.4)\ (\text{ohms } Z)$$

Short-circuit kva of the utility is the maximum short-circuit kva it can produce. Per unit impedance of the utility on its own base kva is 1.0, therefore the per unit impedance of the utility on the chosen base kva is:

$$\frac{(1.0)\ (\text{chosen base kva})}{\text{utility short-circuit kva}}$$

Case Study

Short circuit calculations for a portion of the electrical distribution system are as follows:

A. Utility equivalent impedance. (Values supplied by utility.)

$$°/1Z = \frac{(1.0)\ (10{,}000)}{750{,}000} = 0.0133$$

B. 3,750 kva transformer. (Values from nameplate.)

$$°/1Z = \frac{(0.06)\ (10{,}000)}{3{,}750} = 0.16$$

Above 600 V, the resistance of cables is generally neglected because it has no significant bearing on the total magnitude of short-circuit currents. Below 600 V, however, resistance should be included. Keep in mind that the lower the voltage, the more effective a small impedance is in limiting the short.

C. 520 ft. No. 4/0 single conductor cable. (Value from Table 10–1)

$$\text{Reactance} = \frac{(0.00529)\ (520)}{100} = 0.0275 \text{ ohms}$$

$$°/1X = (0.0275)\ (0.576) = 0.0158$$

D. 1,000 kva transformer. (Values from nameplate)

$$°/1Z = \frac{(0.058)\ (10{,}000)}{1{,}000} = 0.58$$

E. 400 ft. 500,000 cir. mils, single conductor, 480 V cable. (Values from Table 10–2)

$$\text{Reactance} = \frac{(0.00450)\ (400)}{100} = 0.018 \text{ ohms}$$

$$°/1X = (0.018)\ (43.4) = 0.78$$

$$\text{Resistance} = \frac{(0.00359)\ (400)}{100} = 0.0144 \text{ ohms}$$

$$°/1R = (0.0144)\ (43.4) = 0.62$$

The maximum short-circuit current will flow through a circuit breaker, fuse, or motor starter when the fault occurs at its terminals. Therefore, it is not necessary to calculate the currents beyond the last breaker, etc.

Please note in the following calculations that per unit impedances to the first and second faults are merely the arithmetic sum of per unit impedances from the source to the fault points. Since resistance is negligible, reactance and impedance are considered equal. Resistance of the 480 V cable leading to the third fault, however, is not negligible. To calculate per unit impedance at this point, therefore, a vector sum is used. Cable per unit reactance is added to the sum of preceding reactances—assumed equal to impedances since resistances have been neglected. This value and the per unit resistance of the cable are then used to compute per unit impedance to the fault.

Asymmetrical short circuit kva at fault No. 1 is:

$$\frac{10{,}000 \text{ base kva } (1.25)}{0.1733} = 72{,}000 \text{ kva}$$

Symmetrical short circuit current at fault No. 2 is:

$$\frac{10{,}000 \text{ base kva}}{(\sqrt{3})\ (0.48)\ (0.769)} = 15{,}700 \text{ amps}$$

Table 10-1. Approximate Resistance, Reactance and Impedance of 5000 V Cable in Magnetic Duct, per 100 Ft. Three single-conductor cables per duct.

CABLE SIZE	R	X	Z
AWG			
10	0.1240	0.00850	0.1240
8	0.0779	0.00788	0.0781
6	0.0498	0.00748	0.0503
4	0.0318	0.00681	0.0325
2	0.0203	0.00623	0.0212
1/0	0.0131	0.00567	0.0143
2/0	0.0106	0.00545	0.0119
3/0	0.00860	0.00535	0.0101
4/0	0.00700	0.00529	0.00877
CIR MILS			
250,000	0.00609	0.00525	0.00802
300,000	0.00520	0.00519	0.00735
350,000	0.00461	0.00514	0.00690
400,000	0.00419	0.00506	0.00657
500,000	0.00359	0.00495	0.00611
750,000	0.00180	0.00474	0.00551

Table 10-2. Approximate Resistance, Reactance and Impedance of 600 V Cable in Magnetic Duct, per 100 Ft. Three single-conductor cables per duct.

CABLE SIZE	R	X	Z
AWG			
14	0.3135	0.00765	0.3135
12	0.1972	0.00710	0.1972
10	0.1240	0.00687	0.1240
8	0.0779	0.00638	0.0782
6	0.0498	0.00598	0.0500
4	0.0318	0.00551	0.0332
2	0.0203	0.00513	0.0209
1	0.0163	0.00500	0.0171
1/0	0.0131	0.00495	0.0140
2/0	0.0106	0.00490	0.0117
3/0	0.0086	0.00486	0.00986
4/0	0.0070	0.00482	0.00850
CIR MILS			
250,000	0.00608	0.00480	0.00778
300,000	0.00552	0.00474	0.00704
350,000	0.00461	0.00569	0.00658
400,000	0.00419	0.00462	0.00625
500,000	0.00359	0.00450	0.00575
750,000	0,00180	0.00438	0.00520

Asymmetrical fault current is:

$$(1.25)\ (15{,}700) = 19{,}600 \text{ amps}$$

When the connected induction motor load is known, then 3.6 times full load current is added to the above for total asymmetrical fault current:

$$19{,}600 + (2{,}000 \times 3.6) = 26{,}800 \text{ amps}$$

Symmetrical short circuit current at fault No. 3 is:

$$\frac{10{,}000}{(\sqrt{3})\,(0.48)\,(1.68)} = 7{,}150 \text{ amps}$$

Asymmetrical fault current is:

$$(1.25)\ (7{,}150) = 8{,}950 \text{ amps}$$

Including the fault current contribution of the induction motors, the result is:

$$8{,}950 + (2{,}000 \times 3.6) = 16{,}150 \text{ amps}$$

In summation, exact calculation of maximum possible short-circuit currents in a given system is very complicated. However, for all practical purposes this simplified method is sufficient.

CIRCUIT BREAKERS AND FUSES

Circuit Breakers

Circuit breakers are mechanical devices designed to open or close contact members under normal and abnormal conditions. According to the electrical industry, breakers rated for 600 volts or less are considered low voltage.

Circuit breakers are available in any number of different types, categories, frame sizes, and ratings. There are two general types: adjustable and nonadjustable, three general categories; molded-case; solid state tripping; and power air circuit breakers. They are available in a wide range of frame sizes; 100 A, 225 A, 400 A, 600 A, 800 A, 1200 A, 2000 A, 2500 A, 3000 A, and 4000 A. Within these frame sizes are numerous trip settings. Not all manufacturers make all frame sizes and trip settings.

There is continual improvement in circuit breaker technology and the degree of sophistication is constantly expanding to the point where some manufacturer offers one with "microprocessor intelligence." It even offers "energy monitoring option."

Interrupting ratings also vary widely. Most standard CBs manufactured today are rated at 10,000 A minimum, and go as high as 100,000 A at 600 volts, and 200,000 A at 240 volts. Where system short circuit currents exceed interrupting ability of CBs, current limiting fuses are installed on the line side of the CB to protect it from fault currents beyond its design limit. Circuit breakers are designed to provide overcurrent protection for the load and to provide the essential functions necessary for the safety of people and equipment, and to prevent fires and eliminate shock hazards.

For the proper selection and application of circuit breakers, operating conditions that must also be considered include ambient temperature, moisture, dirt, corrosive atmosphere, altitude (low air density at higher altitudes reduces cooling), physical location, vibration, and shock.

Another essential consideration is selective tripping coordination, for application of circuit breakers alone, fuses alone or CBs and fuses combined. The overcurrent or short-circuit protective device nearest the overload or fault only should trip. This isolates the problem and does not shut down more circuits than is absolutely necessary.

TYPICAL SPECIFICATIONS
Circuit Breakers

Air-power circuit breaker, drawout type, 3-pole, single-throw, rated 13,800 volts, 500,000 kva, 18,000 A symmetrical interrupting capacity and 1200 amperes continuous.

Main transformer secondary insulated case drawout circuit breakers, 3 pole, for use in 480-volt switchboard, rated 4000 amperes continuous and 100,000 amperes symmetrical interrupting capacity. Each shall be complete with three solid state, nonderating overcurrent trip devices and a quick make quick break manual stored energy operating mechanism. Trip units shall include long-time, short-time, and instantaneous trip characteristics for coordination with feeder breakers. This circuit breaker shall be complete with static type solid state integral ground fault protection. All trip units shall be adjustable from the front with trip selectors mounted behind transparent windows. The circuit breakers shall be key interlocked with the primary selector switch and the bus tie circuit breaker.

Feeder circuit breakers shall be individually mounted. They shall be manually operated with quick make quick break trip free mechanism, and have three solid state, nonderating overcurrent trip devices. Trip units shall include long time and instantaneous trip characteristics for coordination with the main circuit breakers. Minimum interrupting capacity shall be 100,000 amperes symmetrical. The number, rating, size, and arrangement shall be as shown on the drawings.

Power distribution panelboards shall be dead front circuit breaker type with 3 phase, 4 wire mains. Circuit breakers are to be minimum EH frame, 14,000 AIC. The frame size and rating of circuit breakers are to be as indicated on drawing panel schedules.

Lighting panelboards 277/480 volts and 120/208 volts shall be dead front circuit breaker type with 3-phase, 4-wire mains. All circuit breakers shall be the bolt-on type. Plug-in type will not be accepted. Circuit breakers on 277/480 volts shall have a minimum AIC rating of 14,000 A and on 120/208 volts the minimum AIC rating shall be 10,000 A. (AIC is Amperes Interrupting Capacity)

Fuses

For fuses, unlike circuit breakers, there a only four levels of interest. These are 10,000, 50,000, 100,000, and 200,000 RMS symmetrical amperes of short circuit current. Low voltage fuses can be divided into two categories—0 to 600 A and 601 to 6,000 A. While electrical fuses are deceptively simple protective devices, matching applications in increasingly complex and demanding circuits is becoming more difficult. Fuses can be considered electrical safety valves, carefully chosen to isolate a part of an electrical system when it is in trouble.

Fuses are available in a wide variety of types and classes, a few common types are listed:

1. One-time fuses, with or without time delay
2. One-time fuses with renewable links (reusing the cartridge) without time delay
3. Current-limiting fuses

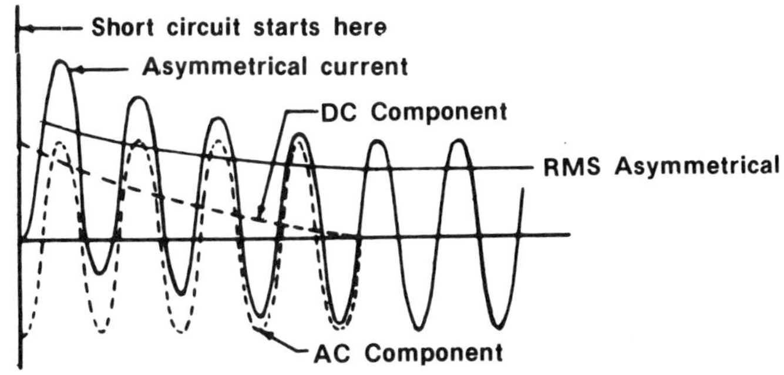

Figure 10-13. A. Fault currents are neither symmetrical or fully asymmetrical, but somewhere in between, depending upon the point on the voltage wave where the short circuit occurs. The DC component decays to zero in about four cycles, the rate of decay (Decrement) depend upon circuit constants.

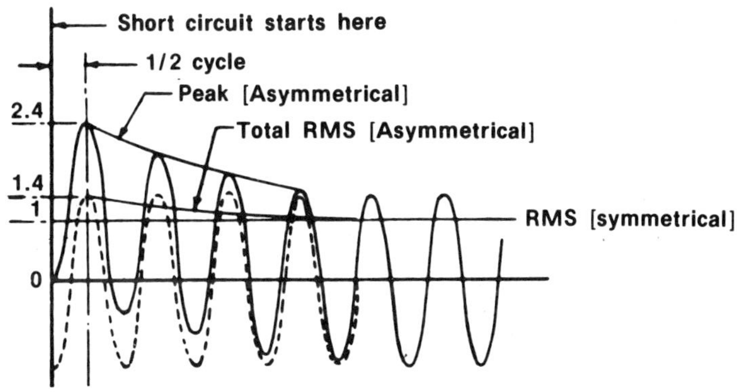

B. Accepted practice is to use the current available 1/2 cycle after the short circuit starts as the available short circuit current. Total RMS asymmetrical current is 1.4 times the RMS symmetrical circuit current, and the peak instantaneous current is 1.7 times the RMS asymmetrical circuit current. (Courtesy of Gould Inc., Electric Fuse Division, Newburyport, MA.)

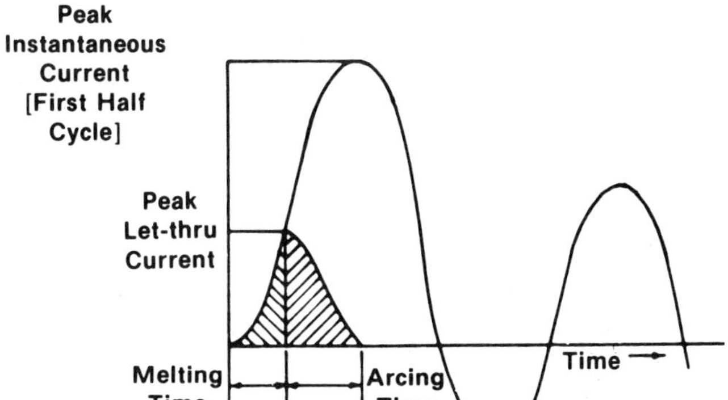

Figure 10-14. Current limiting action of current limiting fuses. *Melting Time* is the time to melt the fusible link. *Arcing Time* is the time required for the arc to burn back the fusible link and reduce the current to zero. *Total Clearing Time* is the sum of the melting and arcing times and is the time from fault initiation to extinction. (Courtesy of Gould Inc., Electric Fuse Division, Newburyport, MA.)

There are several different classes of fuses in common use in this country. Class H fuses are the original cartridge fuses, for use on 250- and 600-volt circuits up to 600 A with 10,000 A interrupting rating. Class K fuses are similar to and interchangeable with Class H, but have a 50,000 A interrupting rating. While Class K fuses are current-limiting, they are generally labeled energy limiting and not current limiting. UL standards and the NEC require that current-limiting fuseholders must not accept noncurrent-limiting fuses.

Class R fuses (RK 1 and RK 5) are labeled current-limiting fuses for use on 250 and 600 volt circuits up to 600 A with a 200,000 interrupting rating. These fuses, by design, cannot be inserted in noncurrent-limiting fuseholders. Class L fuses are for use on 600-volt circuits for currents from over 600 A up to 6,000 A with a 200,000 A interrupting rating. While there are other classes of fuses, the ones mentioned above are more frequently used in power distribution application. Special purpose fuses are used for such applications as for capacitors, rectifiers, welders, cable limiters, etc.

Class R fuses are special purpose fuses dedicated to short-circuit protection. They are not to be used alone for overload protection in any circuit. They are selected to coordinate with overload relays by the motor starter manufacturer. The fuses are chosen with melting time-current characteristics which cross the overload relay curves at 6 to 8 times the full-load current of the motor. The relay handles all overload currents and the current-limiting fuse clears all short-circuit currents.

Current-limiting fuses are designed to reduce the current flowing in a faulted circuit to a magnitude substantially less than that obtainable in the same circuit without a current-limiting fuse.

Short-circuit current reaches its maximum value at the peak of the first $\frac{1}{2}$ cycle. Current limiting fuses operate in less than $\frac{1}{2}$ cycle therefore significantly reduce the available short circuit current. The maximum peak current which pases through the fuse is called the let-thru current. Fuse manufacturers furnish let-thru curves for their various types of current-limiting fuses.

TYPICAL SPECIFICATIONS
Fuses

Fused switches shall be provided with fusetron, low peak Class R fuses and the electrical contractor shall furnish one complete set of space fuses, of each size noted, to the owner.

POWER FACTOR CORRECTION

Power factor correction is a sound economic practice and usually returns the investment in capacitor installation in from one to three years. The rate of return varies widely because of the differences in power factor clauses used by different power companies in various parts of the country.

The most economical and simplest method of power factor correction in existing buildings is with capacitors. They are the simplest of all electrical devices for they contain only metal foil, paper, and an insulating liquid, have no moving parts, consume little power, and require almost no maintenance.

Power factor is simply a name given to the ratio of *actual* power being used in a circuit, expressed in watts or kilowatts (kw), to the power which is *apparently* being drawn from the line, expressed in volt-amperes or kilovolt-amperes (kva).

Power factor may also be expressed as the ratio of power producing current in a circuit to the total current in that circuit. However, the ratio of kw to kva is generally more useful. While the power factor ratio is of great importance in alternating-current circuits, it has no significance in purely direct-current circuits.

$$\text{Power Factor} = \frac{kw}{kva} = \frac{\text{Actual Power}}{\text{Apparent Power}}$$

$$= \frac{\text{volts} \times \text{amperes} \times \text{cosine}\ \Theta}{\text{volts} \times \text{amperes}}$$

$$= \text{cosine}\ \Theta$$

$$kw = kva \times PF$$

$$kva = kw\ /\ PF$$

In a circuit with 40 kw actual power and 50 kva apparent power, the power factor is 40/50 = 0.8; this is commonly expressed as 80 percent.

For load devices containing only resistance, such as electric resistance heaters, the actual power and the apparent a-c power are the same and the power factor is 100 percent. Many devices, such as a-c motors, transformers, and inductors, have a property known as inductance, and these devices consume less actual power than shown by the products of their operating currents and voltages. This inductance in an a-c circuit causes the current to be out of phase with the voltage across that circuit. The greater the inductance in an a-c circuit, the further its current will lag behind the voltage, and the power factor will be lower. In any inductive circuit, the total current in that circuit, as measured by an ammeter, is the vector sum of the power-producing or in-phase current, and the reactive current.

This relationship leads directly to the power factor triangle where:

Actual power = Voltage times in-phase component of the current.
Reactive power = Voltage times reactive current.
Apparent power = Voltage times total current.
(theta) Θ = The angle included between the *kva* and *kw*, and is called the power factor angle. The cosine of this angle (Cos Θ) is the power factor.

All devices containing inductance, such as motors, generators, transformers, and other machinery with coils, require reactive currents to produce the magnetic fields needed for operation. In these circuits, the voltage peak occurs first, and the current is said to "lag" the voltage. In capacitive circuits, the current peak occurs first and the current is said to "lead" the voltage. In purely resistive circuits with no inductance or capacitance, the current and the voltage peaks occur simultaneously and are said to be "in-phase." Here the angle Θ (theta) is always equal to zero.

In circuits containing resistance together with inductance, the lag angle is always

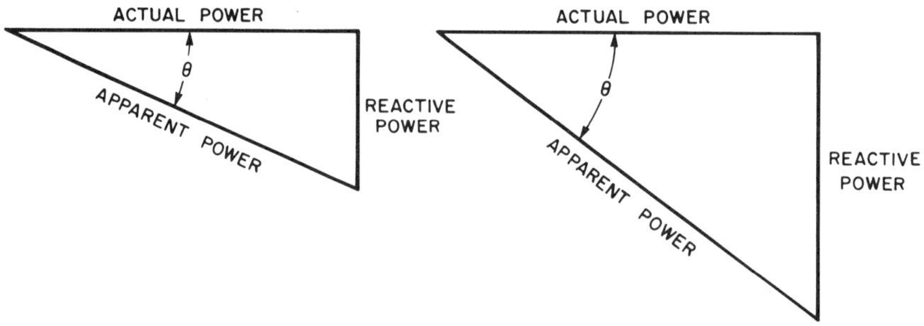

Figure 10-15. When 0 is small, the reactine power is small and the apparent power is almost equal to the actual power. When 0 is large, reactive power is large and the apparent power is much larger than the actual power. (Courtesy of Sprague Electric Company)

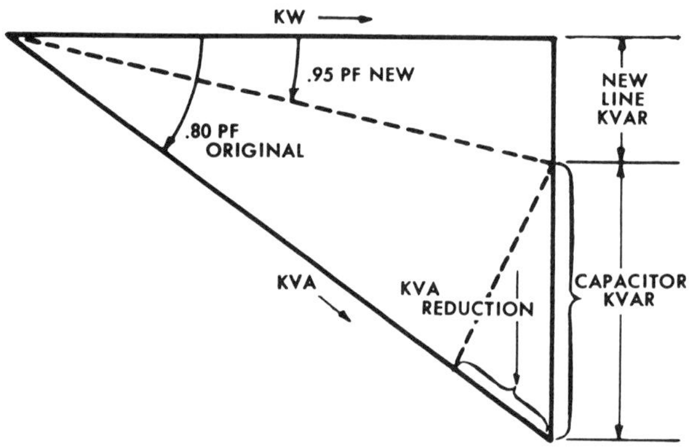

Figure 10-16. Composite power factor triangle showing the relationship of original 80% PF corrected to new 95% PF. (Courtesy of Sprague Electric Company)

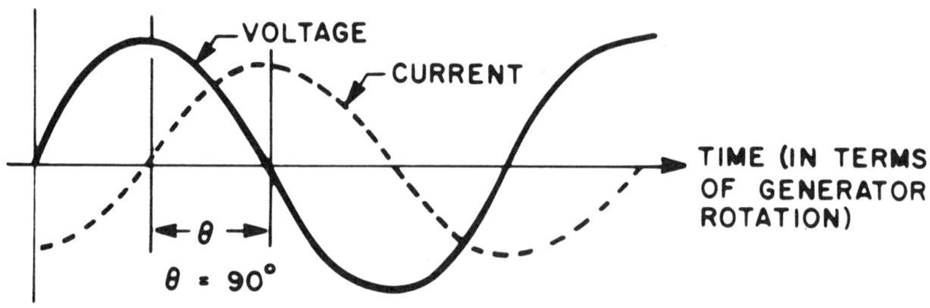

Figure 10-17. In a purely inductive circuit the current lags the voltage by 90°. In actual circuits some resistance is always present; therefore, 0 is always less than 90°. (Courtesy of Sprague Electric Company)

less than 90°. How much less depends upon the relative amounts of each, or more precisely, the $\frac{\text{inductance}}{\text{resistance}}$ ratio. The greater the inductance with respect to the resistance, the greater the lag angle Θ. From a practical standpoint, however, this angle never reaches 90°, because some resistance is always present in every circuit. Since most plant machinery is inductive in nature, the chief concern is with lagging currents.

Major Source of Low Power Factor

The major source of low power factor is the squirrel-cage induction motor. The ruggedness and simplicity of three-phase squirrel-cage motors make them first choice where constant speed is acceptable. For this reason they comprise about 90 percent of all electric motors currently in use in industry.

Power factor for these motors varies from a low of 68 percent for a one hp motor with a synchronous speed of 900 rpm to 90 percent for a 200 hp motor with a speed of 1,800 rpm. These figures are for motors operated at full load. Unfortunately, many applications require that these motors operate at less than full load, i.e., varying load conditions. Conveyors, for instance, are designed for maximum load with motors to suit, but often operate at lesser loads and frequently run unloaded; at this point the power factor is low.

Figure 10–18 is a typical curve for power factor vs motor load. It shows that a 100 hp motor operated at full load will have an 89 percent power factor, but when the load drops to 20 percent it will have a power factor of approximately 50 percent.

For constant load applications, such as pumps, fans, etc., motors should be properly sized in order to obtain maximum power factor. Fortunately, motors usually operate at maximum efficiency at full load.

Note what happens to this same motor when operated at 20 percent load but properly sized capacitors have been applied; the power factor approaches 100 percent.

Effects of Low Power Factor

Low power factor means poor electrical distribution efficiency, which is always costly—because the actual power consumption is less than the volt-ampere product. It is the volt-amperes which the electrical system sees, but it is the watts that the load uses.

With low power factor, the volt-ampere value exceeds the watts consumed. This extra current is drawn through the power lines, transformers, and system distribution lines, or is supplied by the in-plant generator. There are several reasons why low power factor is costly.

1. Low power factor increases the power company's cost of supplying actual power, because more current must be transmitted and this higher cost is billed directly to the industrial consumer by means of power factor clauses in the rate schedule.

2. Low power factor also causes overloaded generators, transformers, and distribution lines within the industrial plant; voltage drops and power losses are greater than they should be. All this represents waste, and needless wear and tear on industrial equipment.

Generators are normally rated in kva. Therefore, when the reactive current needed by induction devices must be supplied by a generator, the generator's useful output is greatly reduced. A reduction in power factor from unity to 80 percent reduces kilowatt output by as much as 27 percent.

Transformers are also rated in kva similar to generators. At a power factor of 60 percent, the available kilowatt power is about 60 percent of its nameplate rating. In

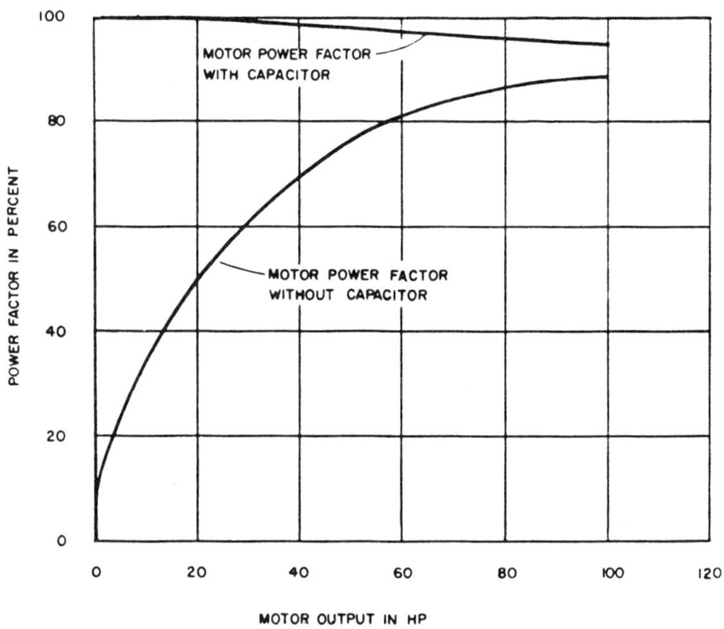

Figure 10-18. Power factor curves for 100 hp motor with and without power factor correction.

addition, the percent regulation is more than doubled between a power factor of 90 percent and a power factor of 60 percent. For instance, a transformer with regulation of 2 percent at 90 percent power factor may increase to 5 percent at 60 percent power factor.

3. Finally, low power factor reduces the load handling capability of the plant electrical system.

Power Factor Clauses

An electric power company must provide the necessary power to all of its consumers. This includes their needs for reactive power, which does not register on the company's kilowatt-hour meter. To supply these needs, a power company must spend extra money to furnish higher rated generators, power lines, transformers, and other equipment on which the company receives no direct return.

Consequently, most power companies make up the added expense by including power factor penalty clauses in their billings. The cost of reactive power, or, as it is commonly known, kilovars consumption, may amount to a significant portion of the monthly bill. Substantial savings can be made by installing capacitors to supply these kilovars.

While power factor clauses perform the same basic function, i.e., to penalize for low power factor, these clauses are not the same for all power companies.

The most common type is the kilowatt-demand where billing demand is dependent upon the ratio of base to actual power factor:

$$\text{Billing kw demand} = \text{Actual kw demand} \ \frac{(\text{Base PF})}{(\text{Actual PF})}$$

A second type is similar to the above, plus an energy (kilowatt-hour) charge dependent upon billing demand. In this case power factor improvement becomes very attractive as both the demand and energy charges are reduced.

Some companies use the flat percentage charge applied to the total monthly bill, depending upon the amount the actual power factor is less than the base power factor.

The first step when considering power factor correction for reducing power costs is to determine if there is a power factor penalty clause imposed by the power company. Where there is a penalty clause (and there usually is), obtain from your power company's representative the charging method. He will gladly supply this information because he, too, is interested in correcting low power factor conditions. Finally, determine the penalty costs for your plant and the corrective capacitor installation costs and then calculate the rate of return. While this rate will vary widely, it usually pays to apply capacitors for power factor correction.

Advantages of Power Factor Correction

In addition to actual savings in power costs as a result of improved power factor, many other benefits can be obtained. When savings in power costs are the only consideration, then the group installation method is more economical. In this case, large banks or racks of capacitors are installed at the main switchboard and the overall power factor is increased, but it should be noted that no benefits to the plant distribution system are obtained on the load side of the capacitor installation. All benefits of capacitor installation are from the capacitor location back to the source of power. The main reason that bank installations are used is that large banks of capacitors have a lower price per kilovar than individual units, and are less expensive to install. However, added benefits can be obtained, and these depend upon the location of the capacitors.

There are three major advantages of improved power factor:

1. Released system capacity
2. Improved voltage conditions
3. Reduced power losses

Improved power factor releases system capacity because reactive current is reduced, and this in turn reduces the total current. As was said earlier, it is the volt-amperes which the system sees.

Transformers, circuit wiring, and generators are limited in capacity by temperature rise, and the heat produced in a conductor is proportional to the square of the current. Therefore, when the current is reduced from 100 to 90 percent, only 81 percent of the heat is produced than before the current was reduced.

While there are numerous curves, charts, and formulas that can be used to determine the amount of system capacity released by power factor correction, the following method is one of the easiest to apply:

$kva\ (\text{Original } PF) - kva\ (\text{Improved } PF) = \text{Released capacity}$

Determine the released system capacity when the power factor is raised from 80 percent to 95 percent with a load of 100 KW.

$$kva_{(80\%)} = \frac{kw}{PF} = \frac{100}{.80} = 125\ kva$$

$$kva_{(95\%)} = \frac{kw}{PF} = \frac{100}{.95} = 105\ kva$$

$$\text{Released capacity} = 125 - 105 = 20\ kva$$

Power factor correction capacitors, installed directly at the terminals of inductive loads such as motors, generate most or all of the reactive power needed to set up the magnetic fields of motors, and thus reduce or eliminate the need to supply this power from the distribution system.

For example, where four motors operate at a power factor of 75 percent, power factor correction to 95 percent will release enough system capacity to accommodate an additional motor of the same size. This means, for instance, that, after correction, five 100 hp motors can be operated on feeders and transformers that previously accommodated only four.

Where transformers and circuits are being overloaded, power capacitors installed at the various sources of inductive reactive loads can release system capacity or relieve the overload to allow service to added equipment. The installation of power capacitors may, in some instances, eliminate the need for installing larger power transformers, rewiring, or possibly both.

Use Figure 10-20 to determine how much system capacity is released when the system power factor is improved.

Problem: A building has a load of 290 kw with an average power factor of 70 percent.

1. How much capacitor kvar is required to increase power factor to 90 percent?
2. How much system kva will be released as a result of improved power factor?

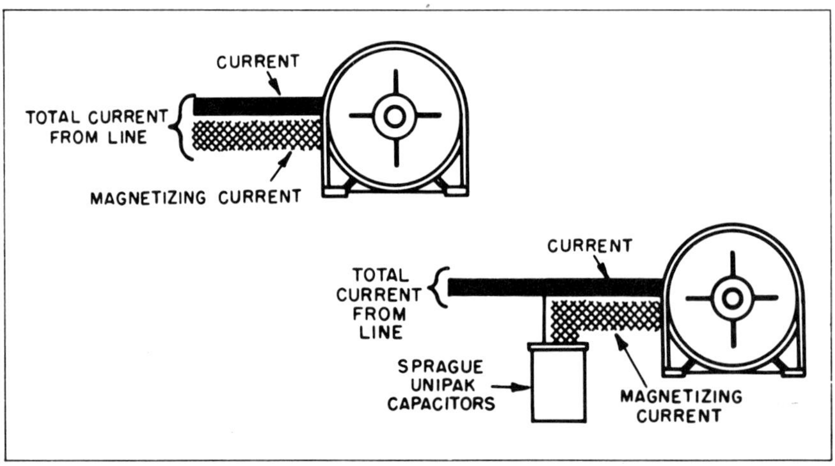

Figure 10-19. A. Induction motor without power factor correction. Feeder line must carry both magnetizing (reactive) current and useful current.
B. Results of capacitor installed at motor terminals. Feeder carries useful current only and capacitor provides magnetizing current. (Courtesy of Sprague Electric Company)

Solution:

1. To find the value of capacitors required, enter left side "Original Power Factor" at 0.70 and extend to intersection of curve labeled 0.90. Drop to base and read 0.54 Capacitor kvar. This reading multiplied by 290 kw results in a capacitor value of 156.6 kvar. Use the nearest standard value, which is 160 kvar.

2. The amount of system capacity released is found by entering the right side "Original Power Factor" line at 0.70 and extending to curve 0.90. Dropping to the base line shows a value of .32 which, when multiplied be 290 kw, results in released kva value of 93 kva. This means that an additional 93 kva of 70 percent PF load may be added to the existing system.

The second advantage of improved power factor is improved voltage conditions. Low power factor can reduce plant voltage when kilovar current is drawn from the feeder line. As power factor decreases, the total current increases (from greater reactive current), causing greater voltage drops across the line resistance. This is because the voltage drop in a line equals the current flowing through the line multiplied by the line resistance. The greater the current, the greater the drop. While capacitors reduce the voltage drops, it is seldom economical to apply capacitors for voltage improvement alone.

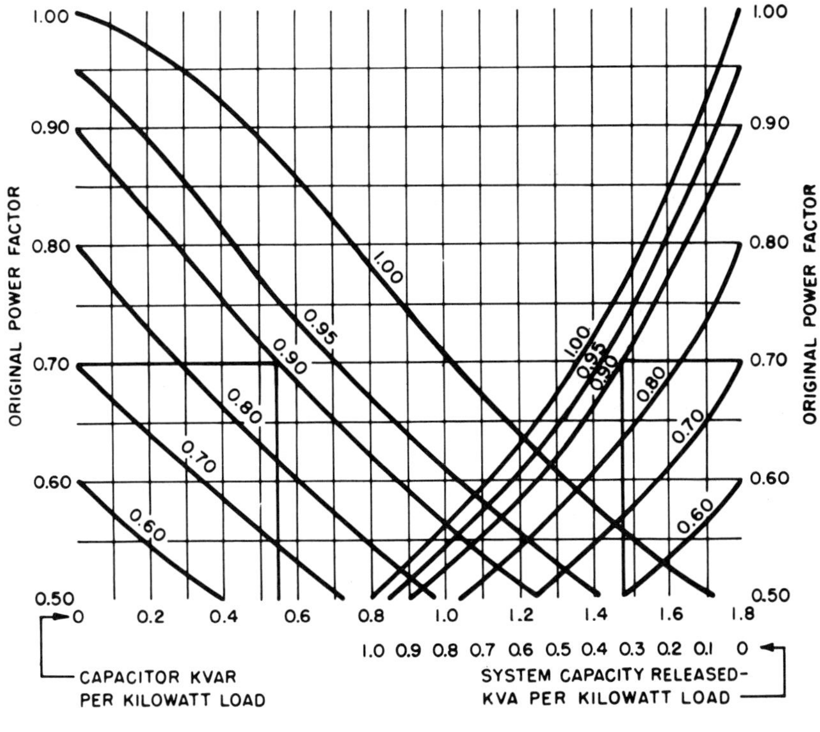

HOW SYSTEM CAPACITY IS RELEASED AS POWER FACTOR IS IMPROVED

Figure 10-20. (Courtesy of Sprague Electric Company)

Capacitor Application

There are two ways in which capacitors can be applied to the system, i.e., shunt or series. The terms "series capacitors" and "shunt capacitors" are general terms denoting the method of connection and really mean series connected capacitors and shunt connected capacitors.

While the major difference between the two is in the way they are connected in the circuit, they do perform different functions. The series capacitor is primarily used to correct severe voltage conditions as encountered with resistance welders, high frequency generators. Series capacitors provide an automatic and instantaneous voltage rise as a function of the circuit current under severe load conditions. Therefore, they are sometimes considered voltage regulators, but they only correct for voltage drops (or voltage dips) imposed on the circuit by loads in series with the capacitors. They cannot compensate for voltage fluctuations originating in the power source.

Series capacitors at rated loads provide power factor improvement by supplying an out-of-phase component of voltage; shunt capacitors provide an out-of-phase component of current. While series capacitors can be used for power factor improvement, they are not used for this purpose because their kilovar rating is too low for significant improvement. They are considered special applications for severe voltage conditions. Since this chapter deals with power factor improvement, the above was intended as an introduction to the series connected capacitor.

Capacitor Installation

There are two general methods for capacitor installation, i.e., group or bank installation and individual installation.

When the chief reason for power factor correction is to obtain lower power bills, or to reduce the current in the primary feeders or the transformer banks, then the group installation is more economical. In this case, large banks or racks of capacitors are installed at the main switchboard. The overall plant power factor is then increased, but it should be remembered that no benefits to the plant distribution system on the load side of the capacitor installation are obtained.

The installation of individual capacitors provides the greatest benefits when wiring is being overloaded by induction motors. This is of more concern in older buildings but also applies to new installations where growth has led to the addition of more motors. Power factor correction can, in such cases, mean the difference between an entirely new wiring job or the simple addition of the new equipment to old lines.

The most effective correction is obtained when the individual capacitors are connected directly to the terminals of the motors and other inductive equipment. Reactive current causes losses either between the induction machine and the capacitor, or, when the capacitor is lacking, between the machine and power source. Thus, the closer the capacitor to the machine, the lower the losses, and the greater the benefits.

Another advantage in placing capacitors directly at the motor is that the motor and capacitor can be switched on and off together. Unlike the group or bank installation method, extra switches are not needed, and the capacitor is switched with the motor.

Flexibility is another advantage of the individual capacitor installation. When the

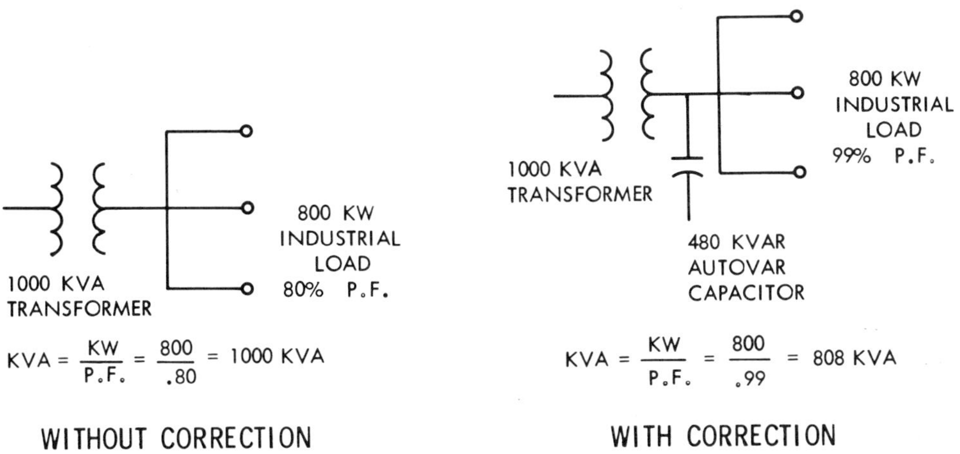

Figure 10-21. A. An 800 KW operating at 80% PF fully loads a 1000 KVA transformer. B. The same load operating at 99% PF imposes 808 KVA on transformer. (Courtesy of Sprague Electric Company)

layout is changed or machinery relocated, both the motor and its capacitor can be moved with minimum inconvenience.

No complicated engineering studies are needed when capacitors and induction motors are connected together and operated as a unit. The size of the capacitor needed for each motor may be determined very simply from Figure 10–22 by looking up the motor speed and horse-power and reading off the required capacitor value in kvar. The table is for use with 3-phase, 60-cycle NEMA Classification B Motors to raise full-load power factor to approximately 95 percent.

The chief disadvantage of individual capacitor correction is that the smaller capacitor units have a higher price per kilovar than larger units. When many small motors on the same circuit need to be corrected it may be more economical to correct them in groups, using 10 kvar to 15 kvar rated capacitors connected to the feeder lines. A common practice is to correct motors larger than 10 hp with individual capacitors, and to correct smaller motors in groups.

Capacitor Installation Considerations

Capacitors provide a reliable, effective, economical means for power factor correction. They have a long service life and require almost no maintenance. But to get maximum benefits out of capacitors, proper installation is important. Among points to be considered are ambient temperature, adequate ventilation, and dust free atmosphere.

Capacitors should be located where the temperature of the air does not exceed 104°F. They should not be installed in small rooms near radiators or other types of heating units, or in outdoor areas (unless designed for outdoor use) where the direct rays of the sun might raise the capacitor temperature excessively.

Capacitors should have adequate ventilation, since they always operate at full load and generate heat which must be removed for long life operation.

Dust accumulation restricts the free flow of air required for normal ventilation; therefore, dust laden areas should be avoided.

SUGGESTED MAXIMUM CAPACITOR RATING WHEN MOTOR* AND CAPACITOR ARE SWITCHED AS UNIT

Induction Motor Horse-Power Rating	Nominal Motor Speed in RPM											
	3600		1800		1200		900		720		600	
	Capacitor Rating KVAR	Line Current Reduction %	Capacitor Rating KVAR	Line Current Reduction %	Capacitor Rating KVAR	Line Current Reduction %	Capacitor Rating KVAR	Line Current Reduction %	Capacitor Rating KVAR	Line Current Reduction %	Capacitor Rating KVAR	Line Current Reduction %
3	1.5	14	1.5	15	1.5	20	2	27	2.5	35	3.5	41
5	2	12	2	13	2	17	3	25	4	32	4.5	37
7½	2.5	11	2.5	12	3	15	4	22	5.5	30	6	34
10	3	10	3	11	3.5	14	5	21	6.5	27	7.5	31
15	4	9	4	10	5	13	6.5	18	8	23	9.5	27
20	5	9	5	10	6.5	12	7.5	16	9	21	12	25
25	6	9	6	10	7.5	11	9	15	11	20	14	23
30	7	8	7	9	9	11	10	14	12	18	16	22
40	9	8	9	9	11	10	12	13	15	16	20	20
50	12	8	11	9	13	10	15	12	19	15	24	19
60	14	8	14	8	15	10	18	11	22	15	27	19
75	17	8	16	8	18	10	21	10	26	14	32.5	18
100	22	8	21	8	25	9	27	10	32.5	13	40	17
125	27	8	26	8	30	9	32.5	10	40	13	47.5	16
150	32.5	8	30	8	35	9	37.5	10	47.5	12	52.5	15
200	40	8	37.5	8	42.5	9	47.5	10	60	12	65	14
250	50	8	45	7	52.5	8	57.5	9	70	11	77.5	13
300	57.5	8	52.5	7	60	8	65	9	80	11	87.5	12
350	65	8	60	7	67.5	8	75	9	87.5	10	95	11
400	70	8	65	6	75	8	85	9	95	10	105	11
450	75	8	67.5	6	80	8	92.5		100	9	110	11
500	77.5	8	72.5	6	82.5	8	97.5	9	107.5	9	115	10

*For use with 3-phase, 60 cycle NEMA Classification B Motors to raise full load power factor to approximately 95%.

Figure 10-22. (Courtesy of Sprague Electric Company)

SUMMARY

1. The primary function of an electrical distribution system is to provide power to some terminal (utilization) equipment in a safe, reliable, and economical manner.

2. Transformers are used to transform voltages (and currents) from one level to another. There is no simpler device to perform this function. At the generating station the voltage is stepped up to highest practical value for economical transmission. This voltage is then progressively stepped down to lower levels for local transmission. Finally, at buildings, it is stepped down to utilization values, generally below 600 volts. High-voltage transmission is economical because low values of current are carried by the transmission conductors. At the user end, insulating and terminating high voltage conductors create imponderable problems and may create extreme hazards. Low-voltage (below 600 V) insulation and termination is economical and essentially eliminates extreme hazard.

3. Transformer secondary voltage selection should be the highest practical voltage consistent with the needs of the loads.

4. While the NEC establishes minimum conductor sizes, larger sizes may have to be selected to limit voltage drop.

5. Motor selection depends upon the application and motor characteristics must match load requirements. Excessive overloads and inappropriate ambient conditions have deleterious effects upon both winding insulation and bearings, and will significantly reduce their life.

6. In work requiring the services of electrical engineering consultants, they will either perform short-circuit calculations or specify that the equipment manufacturers do it. In these cases, your involvement is limited to knowing that short circuits must be adequately protected.

7. Circuit breakers and fuses must be replaced with like type only to preserve design protection.

8. Power factor correction is sound economic practice. It should be provided in new construction and should be considered in existing buildings. Economical evaluation will readily determine benefits and establish return-on-investment. If you lack the appropriate expertise, electrical consultants, utility company representatives or capacitor manufacturers can provide the necessary economic evaluation.

9. A well-designed electrical system will provide safe and reliable power distribution. Manufacturer's maintenance recommendations must be followed to preserve the integrity of the system and to assure anticipated life of the equipment.

10. Electrical systems and equipment are a means to an end. With the exception of lighting and office equipment, it provides power to mechanical equipment.

11. Move and Occupancy

Move and occupancy is the culmination of a project that could have started with a site search, progressed through the various design phases by an A/E firm and survived the agonies of construction. It could bring to fruition a project of four to five years duration. It could be a simple, pleasant, and rewarding experience that leaves everyone involved with a sense of accomplishment and a great deal of satisfaction. Or it could be a frustrating experience, frought with seemingly imponderable problems.

Move and occupancy could be as simple as a move into a 1000-square-foot leased office space or as complex as a move to a newly constructed corporate headquarters 500,000-square-foot high-rise building. In either case, permission to occupy is needed before the move can be accomplished. With the leased space, this is merely a line or two on the lease agreement, stating that the tenant may occupy the space on a certain date. For a newly constructed building, the owner requires a Certificate Of Occupancy. This, sometimes, can be a significant obstacle. But more about the C of O later. The preparation of the space is also different; in the former case, the landlord usually has that responsibility, whereas in the latter, the owner/occupant carries the burden. Also, the maintenance of the building and equipment in a leasing situation rests with the landlord, while in a owner-occupied building the owner must prepare for and actually perform the maintenance. Finally, space planning for a small leased space may consist of a freehand sketch showing placement of furniture, and if for some reason the specific placement isn't satisfactory after occupancy, it frequently isn't much work to make the necessary adjustment. On the other hand, space planning for a large building is a highly complex project frequently demanding the services of an experienced professional.

Suffice it to say that there are similarities with and differences between the two extremes cited. Planning and preparation and the actual execution of move and occupancy requires the essential attention consistent with the scope of the project. Since the end result of a leased space or new building construction is occupancy, that must be the underlying consideration from the very inception of the project through to the actual move to occupancy. While this may sound like a simplistic statement, it isn't, so don't become complacent; plan and prepare for move and occupancy because occupancy is the ultimate goal. The entire project becomes an exercise in futility if the space cannot be occupied.

PLANNING AND PREPARATION

Move and occupancy actually consists of two phases; planning and preparation, and the physical move. Planning often starts before the building design phase and usually must be updated continually throughout the construction phase to reflect current requirements. The degree of involvement and the extent of planning depends upon the size of the project and the experience of the owner. In order to address the numerous considerations for the proper planning phase of move and occupancy, items will be listed as required for a large building to be occupied by an owner without any prior experience.

This checklist is intended as a guide and includes those items normally encountered but there may be other specific requirements unique to a particular type of project. Each item will be discussed individually and where this material is covered elsewhere in this book, appropriate reference will be included.

PLANNING AND PREPARATION CHECKLIST

1. Furniture standards
2. Space standards
3. Space utilization
4. Furniture layout
5. Receptacle and telephone outlets
6. Special electronic terminals
7. Finish schedule (materials and colors)
8. Graphics
9. Construction completion schedule
   (By floor and/or area)
   a. Ceilings
   b. Lighting
   c. Painting and wall covering
   d. Floor covering (carpet, tile)
   e. Telephone installation
10. Move contract
11. Occupancy schedule
12. Certificate of occupancy
    (Inspection, test, and approvals by the authority having jurisdiction)
    a. Fire protection system
    b. Smoke evacuation
    c. Fire pump performance
    d. Emergency generator performance
    e. Food service
    f. Electrical installation
    g. Plumbing installation

While the actual move into and the occupancy of a new space occurs after the construction phase of a project is completed, the degree of success of the move and occupancy depends to a great extent upon the thoroughness of planning and preparation. A carefully executed planning stage will significantly enhance the actual move operation.

For example, on a recent 400,000-square-foot four-story office building the move and occupancy was executed without any problem. Considering the fact that the general contractor had some 150 tradespeople in the building at the start of the move, it was considered highly successful. Actually, the move could have been accomplished without assistance because the planning was thorough and appropriate to the size of the project. In addition to the building construction tradespeople, the owner had contracted with and had the responsibility for three separate installation contracts. The planning anticipated this and the scheduling included it.

The general contractor's work was programmed to permit a phased completion so that the owner's separately contracted work could follow that of the building construction. The work in this large four-story building was scheduled for finishing starting at the north end of the fourth story and proceeding south, followed by the third, the second, and the first, then out the door. Ceiling installers started at the north end, followed by the electricians installing lighting fixtures, next came wall covering and painting. Lastly, the electricians trimmed out the receptacles and switches.

When at least 25 percent of this work was completed, the owner's carpet installers started their work. As they proceeded across the floor, the furniture started to arrive. Since the owner purchased all new furniture for this project, the furniture contract called for delivery, uncrating, field assembly, and placement. While this work was proceeding as planned, the telephone installers connected and placed the instruments.

The furniture delivery was scheduled for four one-week periods with a one-week break after the second week to accommodate any unforeseen disruptions. The main ingredient in this operation was early stage planning. Before the furniture contract could be bid, the architect had to complete the furniture layout plan and it had to be approved by the owner. Before the furniture plan could be started, the floor plan had to be completed and also approved. Since the general contractor's work included all interior partitions, the floor plans had to be completed prior to construction contract bidding. Also, before it could be determined that 400,000-square foot of space was required, the owner needed to know how many people were intended to occupy the space, and the amount of space and furniture that were required by the various classifications of personnel.

From this it should be readily apparent why items 1 and 2 are included in the checklist. While the size of a building can be determined and space utilization drawings can be completed without furniture and space standards, there are several significant advantages to having them.

Furniture Standards

Chapter 2 addresses furniture standards. The major reason for having a furniture standards program is to create uniformity. Once a work function has been established for a particular position, a workstation can be developed based upon the type and quantity of furniture required to perform that specific task. This workstation becomes the basis for determining the amount of space required by the department by tabulating individual workstations. Space utilization drawings are then developed using these departmental requirements in the form of "block" diagrams.

Furniture standards are used throughout the entire corporation for uniformity and consistency. Furniture standards are also used for developing cost estimates. For example, once individual workstation furniture costs are established, the total cost is the sum of these work stations.

Example

FURNITURE GROUP	COST PER GROUP	NUMBER OF GROUPS	TOTAL COST
A	$3,500	1	$ 3,500
B	2,800	3	8,400
C	2,100	10	21,000
D	1,800	20	36,000
		Total Cost	$68,900

Furniture standards can be used for obtaining purchasing cost quotations.

Space Standards

Chapter 2 addresses space standards. Space standardization is essential in order to maximize space utilization. Most all major corporations have both furniture and space standards. No matter whether the project is a simple leasing situation or an extensive new building construction, the use of established space standards can save considerable time and money in the space planning phase. The amount of time required by an A/E firm can be significantly reduced and the concept of space planning can be simplified. Also, the owner's participation time can be reduced. Furniture and space standards should be developed for all but the smallest of companies. Once developed, these standards can be very easily and simply updated to reflect changes in operating philosophy.

Space Utilization

In some cases, architects combine the space utilization (block diagram) drawing with the furniture layout drawing. More frequently, the block diagram drawing is developed to show departmental juxtapositioning and specific locations. Block diagramming is an effective method for quickly and efficiently locating blocks of space to determine interdepartmental traffic patterns. After all the blocks are placed in the most appropriate locations for exterior contact, i.e., purchasing and personnel adjacent to the main entrance, to limit this unique traffic for minimum intrusion into normal daily operations, and shipping and receiving at the dock area, the preliminary size and shape of the space can be determined.

These blocks are then filled in using space standards for the various work stations. This is the final test for the block diagram drawing as it determines the functionality of the space.

Furniture Layout

If the space and furniture standards have been adequately developed, there should be no problem completing the furniture layout drawing. This is essential to the move and occupancy phase. This drawing can also be used for telephone equipment installation, workstation identification by occupant's name and for furniture placement within individual spaces. When new furniture is involved, individual pieces are placed by size and type identification; for example, any 30 × 60-inch desk can be placed in any space requiring a 30 × 60-inch desk. In this case, the furniture manufacturer will use this drawing for scheduling shipments to suit construction completion. When the project calls for a move from an existing building, all furniture must be uniquely identified.

Receptacles and Telephone Outlets

Receptacles and telephone outlet drawings frequently do not include the furniture layout. In that case, these two drawings should be "overlaid" to verify that receptacles and telephone outlets are properly located. While these drawings are not necessary for move and occupancy, they are required as part of building construction.

Special Electronic Terminals

More and more office functions are performed on electronic devices; therefore, these must be considered for both construction purposes and equipment placement. This information should be superimposed on the furniture drawings to verify equipment locations and to expedite equipment placement during the move operation.

Finish Schedule (Materials and Colors)

Interior designers usually select finish materials and color schemes. This is important to the move phase only if the owner intends to contract it separately, for example, wall covering, carpet installation, etc., then these drawings are used for bidding and scheduling this work.

Graphics

Graphics drawings become part of move considerations in those cases only when graphics are to be part of the move contract.

Construction Completion Schedule

The most significant problem encountered during the move and occupancy phase is the result of an unrealistic construction completion schedule. In those cases, particularly where the building is large and new furniture is involved, or where an existing lease termination is fixed, problems could be tremendous. Therefore, construction schedules must be thoroughly evaluated and constantly monitored.

In the event a schedule slippage is noted during the construction phase, it must be addressed immediately. There are several courses of action available to the owner depending upon the cause of the schedule slippage. All general construction contracts have special clauses that define the contractor's responsibility for maintaining the approved construction schedule and there are other clauses that describe those delays which are excluded from the contractor's responsibility. For these delays, for example, strikes, acts of God, owner's changes, A/E errors and omissions, the contractor is awarded an extension of the contract.

For delays beyond the responsibility of the contractor, the owner has two options:

1. Agree to compensate the contractor for added effort, such as overtime, increased manpower loading, etc., to recover the anticipated lost time, or
2. Accept the delay. The owner must then reschedule the separate contracts including the move. Where existing lease expiration is involved, this must be evaluated to determine its impact.

Mostly, this decision is simply a matter of cost. If both options are available and can be implemented, the one that increases the construction cost the least is usually selected.

In cases where the contractor is at fault, the owner must enforce the general conditions of the contract.

The importance of establishing a realistic construction schedule cannot be overemphasized. It is an essential consideration of any construction project. Secondly, this schedule must be reviewed and monitored at regular and frequent intervals to insure compliance with the schedule. As the construction nears completion, the schedule established by floor and area must also be reviewed for suitability of the move and occupancy. Contractors should be able to determine the average amount of ceilings, lights, floor covering, etc., that can be installed daily or weekly and this standard is then used to monitor final completion.

1. Ceilings—Most buildings use some type of suspended ceiling system, it could be one with an independent suspension system and lay-in ceiling tile and lighting fixtures, or an integrated system where the lighting fixtures are part of the suspension system and only tile is fitted in later. After all equipment and systems, above the ceiling, are installed, inspected, and tested, the ceiling installation is usually the first step in space finish.
2. Lighting—Lighting installation naturally follows the ceiling and is usually performed by a separate trade. Once the lighting installation is complete, there will be light for all the following work.
3. Painting and Wall Covering—In some instances this is done before the ceiling depending upon the interface between the ceiling and wall. Frequently it is performed after there is sufficient light to insure a quality application.
4. Floor Covering (Carpet, Tile)—Carpet or tile is installed after all overhead work is completed in most cases. However there are times, when, because of scheduling, particularly floor tile, it may be done out of sequence.
5. Telephone Installation—This depends entirely upon the local phone company installation procedure. Before the move can be accomplished at least all wires should be in place ready for instrument attachment. On most projects they will be more concerned about completion within the proposed schedule than you. Just to insure successful occupancy, you should make sure that the final portion of phone work undone at the start of the move is limited to instrument attachment. Just remember, without phones most company operations are unable to function.

Move Contract

The move contract must be settled as soon as possible during the early stages of construction in order to insure a successful move. The contractor selected must be one that has specific experience in the particular type and scope of move. It is to the owner's benefit to consider only contractors capable of executing the move successfully and within the established schedule. Some large companies have movers whose past performance was considered satisfactory and frequently use them repeatedly. The major consideration here is to try to establish the move contract as early as possible and to maintain contact with the mover as the move date approaches.

In cases where this mover has never worked for you before, the following is a list of items of information that the mover should either provide on request, and/or matters that should be discussed at an initial meeting:

1. Inventory of all equipment involved in the move.
2. Make a visual inspection of all the equipment with particular attention to physical size and weight.
3. Check clearances of doorways, elevators, stairways, passageways, etc.
4. Definition of responsibility for protecting floors, walls, elevators, etc.
5. Visit new site and establish furniture movement routing.
6. Determine adequacy of move supervision and coordinators.
7. Discuss cost, use, and quantity of cartons or tote boxes required.
8. Determine who, mover or owner, packs and unpacks which types of material.
9. Establish responsibility for arrangement for traffic flow and use of curb space for trucks.
10. Establish hourly rates for labor and trucks. (Be sure mover hires only experienced employees and not casual labor)
11. Discuss insurance coverage.

Occupancy Schedule (By Floor and/or Area)

In all but the smallest projects, it is usually necessary to develop an occupancy schedule and, at times, this matches the move schedule. On those projects where new furniture and equipment is involved, the entire move is frequently completed before occupancy by personnel. Where existing furniture is reused, the personnel occupying the space and using the furniture must naturally be moved at the same time. In this case, it is important to move related functions in such a manner that minimum operational disruption occurs. The owner must determine his own operational method and plan the move accordingly.

Certificate of Occupancy

A certificate of occupancy is required before the building may be occupied legally. While the owner's role is minor in most cases, he should be an aggressive participant in planning and seeking the C of O. It is issued only after various portions of the work are inspected and approved by the different authorities having jurisdiction.

For example, the local fire marshall must approve all related fire protection systems and equipment. While the fire protection system may have been designed to satisfy all code requirements, the local fire marshall may have unique requirements depending upon his fire fighting equipment and methods that must be addressed and satisfied.

Further, the fire marshall will witness all tests on equipment and systems relating to fire protection, such as, smoke evacuation, fire pump performance, emergency generator performance, location of emergency control stations including elevator controls, placement of fire hose cabinets, and length of fire hoses provided, kitchen equipment fire extinguishment system, etc.

Food service installation must be inspected and approved by the local authority having jurisdiction. In order to insure compliance with its particular requirements, drawings may be submitted to it, or meetings held, as soon as practicable. Final inspection and approval is then greatly enhanced.

Plumbing and electrical systems, equipment, and installations must be inspected and approved, as with the others above, before the C of O will be issued. Plumbing and electrical installations are inspected and tested at various stages of completion as required by the various associated codes.

As stated earlier, the owner's participation is minimal in most cases. The contractor and subcontractors carry full responsibility as described in the typical General Condition of the contract between the owner and the contractor.

TYPICAL SPECIFICATION—GENERAL CONDITIONS

The contractor shall secure and pay for all permits, governmental fees and licenses necessary for the proper execution and completion of the work.

If the contract documents, laws, ordinances, rules, regulations, or orders of any public authority having jurisdiction require any work to be inspected, tested, or approved, the contractor shall make all necessary arrangements for and bear all costs of such inspections, tests, and approvals unless otherwise provided.

The contractor shall secure all required certificates of inspection, testing, or approvals and promptly forward to the consultant.

Some specifications also contain an additional clause:

CLOSEOUT SUBMITTALS

When the work is complete, submit evidence of compliance with requirement of governing authorities as follows:

1. Certificate of occupancy
2. Certificate of inspection for:
   a. Elevators
   b. Mechanical work
   c. Electrical work

Whatever information the contractor needs from the owner, he must be able to obtain it in a timely fashion in order to satisfy his contractual obligations. This is usually limited to actual total construction cost, building area in square feet and number of personnel to occupy the building.

Now the building is complete and is ready for the physical move to the new quarters.

PHYSICAL MOVE

Most experienced movers have standard moving procedures that can be tailored to suit the conditions of any moving situation. In some cases, final details are worked out with the owner's participation and, in some instances, the mover will determine what is required for an efficient and successful move. In either event, the mover will need to tour both the existing and new facilities to determine the exact routing of the moves.

As with the planning and preparation phase, a checklist will serve as a guide to the entire operation. It covers the normal items necessary for a large move, relocating equipment and personnel from an existing building to the new facility.

PHYSICAL MOVE CHECKLIST

1. Economic considerations
2. Scheduling facilities
3. Supervision
4. Packing
5. Marking
6. Move
7. Post move

The success of any move depends to a great extent upon the experience and qualification of the mover, the owner's experience and participation, planning and scheduling of elevators, docks, etc., and supervision capability.

Economic Considerations

Owners frequently purchase all new furniture for the new building because furniture replacement costs are the least at the time of the move. First, existing furniture move costs are saved and secondly, existing furniture has some salvage value even when sold wholesale. This is a particularly important consideration when existing furniture is old and near future replacement is a good probability.

The type of and distance between the two facilities requires careful economic evaluation. For example, the cost of a move from New York to Atlanta is significantly greater than one that takes place within a suburban, low-rise office park. Also, moving from one high-rise building to another high-rise in an inner-city location has a significant impact on costs. Economic consideration is simply a matter of dollars and cents. Evaluate the cost of new furniture against the cost of moving existing furniture plus the salvage value of the existing furniture.

Another economic option at move time is partial replacement, partial repair, and refinish and partial salvage sale of furniture. Sometimes, depending upon age and condition of existing furniture, it may be most economical to replace the oldest with new, and repair and refinish the more recently purchased equipment. In this case, scheduling and phasing the move becomes very critical. New furniture should be purchased in groups by department so the entire functions may move intact with minimum disruption. Existing furniture planned for repair and refinish should be shipped to a local vendor for that service and then moved into the new facility. All old furniture scheduled for sale should remain unmoved and sold from the existing building. Both repair and refinish operation and salvage sale must be contracted for well in advance of the anticipated move date to eliminate unforeseen last-minute disruption of plans.

Scheduling Facilities

This is most important in high-rise or multi-tenant buildings where dock facilities and elevators are not under total control of owner making the move. In new building construction, it is not uncommon to have construction work continuing during the move phase; therefore, it is necessary to consider the contractor's needs for elevators and dock space. In inner-city moves, street and alley traffic could severely disrupt the move operation unless appropriate measures are taken.

Last, but most certainly not least, not every piece of equipment will fit into every elevator and through every door. Many a move has been disrupted because adequate passage was not prepared for some piece of equipment.

Supervision

The owner must provide continuous supervision during the entire move operation at both locations to make decisions when some unforeseen problem arises. Any disruption of the planned move can be quite costly.

Some companies assign a move supervisor who then selects subordinate supervisors (coordinators), one at each location, from the various departments whose responsibility will be to see that everything scheduled to be moved from the existing building is appropriately handled and to direct the flow of equipment into the new building.

On large moves, a central information headquarters must be established at both locations and should contain at least a desk, chair and telephone. In buildings with a central paging system, the command post should be set up at that location. Telephone service should be maintained at these locations during the entire period of the move.

Packing

Moving is a costly operation and it is important to avoid the expense of moving unnecessary weight and bulk. The first step, therefore, is to houseclean all desks, files, cabinets, closets, etc., and discard all papers and other items that are no longer of value. The next step is the physical packing.

The following moving procedure was developed for preparing for and executing a move. Obviously, all situations cannot be mentioned and, therefore, good judgment must be exercised in handling situations not specifically listed. The intent of this procedure is to serve as a guide and may be used to check against the one used by your mover. The mover usually supplies cartons to be used for packing purposes:

1. Bookcases:	Empty—pack contents in cartons.
2. Desks:	Empty—pack contents in cartons. Drawers should be closed and locked (or taped shut) and keys retained in a safe place.
3. File Cabinets:	The contents of filing cabinets are NOT to be removed. The follower should be pulled up tight in all file drawers. Empty file drawers may be used to pack similar files from desks, and telephone books, stationery, paper, books, etc. All files capable of being locked should be locked. On others, the drawers should be taped shut. (Tape only furniture that will not be damaged or marred by the tape.)
4. Storage Cabinets:	Metal storage cabinets must be emptied and contents placed in cartons.
5. Cartons:	Cartons should be filled but not to overflowing. Close by interlocking four top flaps. Each carton to be labeled as instructed.

Test equipment, heavy equipment, office machines, and miscellaneous items require special attention and the directions of the mover must be followed to allow safe transportation and to be insured. The following disclaimer appeared in several movers procedures: "All property belonging to personnel should be removed by the employees. Neither the company nor the moving contractor can be held responsible for loss or damage to employee property."

Marking

The key to a successful move is appropriate marking and color coding. This is particularly important where departmental rearrangement is accomplished during the move. If the move is into a multistory building, a different color could be assigned to each floor. If the building is one story, different colors may be used for each department.

The furniture layout floor plan comes into use at this stage. Each piece of furniture on the drawing must be numbered, and each piece of furniture to be moved into the space must have a corresponding number. Sometimes the floor plans are cut up by individual spaces, such as offices, open areas, conference rooms, etc., and these cut up pieces are taped to doors, columns etc., to show furniture placement.

For similar pieces of furniture, such as desks, files, etc., the colored, numbered labels should be placed on the same location. For example, on desks and tables affix colored label on top right front.

One moving company uses colored tape, appropriately placed, starting at the loading dock and leading directly to the area or floor designated by that color. All the movers had to do was follow the color matching the item being moved.

Move

The degree of success of the physical move is totally dependent upon the effort and thoroughness of preparation to this stage. The best thing for you, the owner, to do is get out of the way of the movers. There is not much you can or should be expected to do if you have accomplished your responsibilities thoroughly and effectively. This is a physical operation that can be expediously performed by an experienced professional moving company. You should be able to sit back and merely oversee your move supervisor and his subordinate coordinators. Most problems that occur at this stage are the result of incomplete or inept planning and preparation.

Post Move

This depends upon the type of contract established with the moving company. In most cases, the employee is responsible for unpacking his cartons and placing that material into its appropriate location. Most moving companies retain ownership of all cartons and will want them broken down and returned to some central location for retrieval.

SUMMARY

Move and occupancy needn't be a traumatic experience. This final phase of a project, whether it be leased space or new building construction, could be a satisfying experience and a successful completion of a long drawn out project. A simple step-by-step procedure should be developed that addresses all the essential considerations for the move and occupancy phase. The following is a guide that could be used to develop one appropriate to the project at hand:

1. Determine as early as possible, by suitable economic evaluation, the optimum choice or combination of new furniture purchase/refurbishment/disposal.
2. Review furniture and equipment layout drawings to make sure that they are complete and truly reflect the new space occupancy requirements.
3. Select and obtain a contract with a moving company that can best satisfy the requirements of project and the move schedule.
4. Check to verify that all doors and passageways are adequate for the sizes involved.
5. Monitor construction progress and compare with the construction schedule.
6. At least three months prior to completion of construction, check status of certificate of occupancy, and continually monitor progress.
7. Assign supervisor to oversee physical move.
8. Set up command post to monitor and coordinate move, and check post move operation.
9. Close out project.
10. Heave a sigh of relief now that a project which could have started with a site search, progressed through design and ended with a successful move and occupancy.

Index

Index

U

V

W